Wind Turbine Design Simplified

Aerodynamics

Paul L. Gay

Copyright © 2019 Paul L. Gay
Westport MA.   All rights reserved.

ISBN #: 978-1-71600-001-0

About the Author

The author Paul L. Gay has owned wind turbine manufacturing companies and has been responsible for the design and development of both variable speed and stall regulated wind turbines and towers. He designed and patented aerodynamic rotor blade tip controls. He has designed and developed wind turbine electronic control systems using programmable microprocessors. His designs have included folding towers both guyed and freestanding which allows turbines to be lowered to the ground for installation and maintenance.

Paul has taught and developed a wind turbine technology course at the college level under a grant from the National Science Foundation. He was the subject of a 2009 article *Modelling Innovations Advance Wind Energy Industry* by the NASA Scientific and Technical Information publication in which his wind turbine designs and innovations were discussed.

Paul is licensed as an attorney and land surveyor and has published several books some of which are listed below.

Other Books by the Author

Practical Boundary Surveying, Springer International Publishing, 2015.

Land Surveying Simplified, Tupelo Press, 2016.

Land Surveying Mathematics Simplified, Tupelo Press, 2018.

A History of Gray's Mill, Tupelo Press, 2016.

Fundamentals of Boundary Surveying, Professional Surveyor Publishing, 2002.

Preface

This book on wind turbine aerodynamics is the first book in a series of books on wind power by the author. The books are an attempt to present a simplified explanation of wind power technology without sacrificing an in-depth understanding of the subject matter. A number of excellent books on wind power already exist. Books such as W*ind Energy Explained* (Manwell, McGowan & Rogers, 2011) and *Wind Energy Handbook* (Burton, Sharpe, Jenkins & Bossanyi (2004) provide an exhaustive treatment of the subject. However these books are extremely technical. The books contain a relatively high level of mathematics which requires the reader to possesses the skills of a mathematician, scientist or engineer. While these books may be suitable for a college level, multi-semester course on wind power the prerequisites may intimidate many who might want to acquire an understanding of the technology but may not have the requisite mathematical background.

Many other books on wind power exist at a level which is essentially devoid of mathematics. These books are descriptive in nature. Books at this level may appeal to a much broader audience, but they fall short of being suitable for those wishing to acquire a well-grounded understanding of wind power technology. Acquiring even a moderately comprehensive understanding of wind power requires the reader to become familiar with many disciplines. It is necessary to have a familiarity with aerodynamics, electricity, electronics, mechanics, structures and atmospheric science, to name but a few. A book which only provides a description of these subjects, may be interesting reading, however, it will not provide the reader with an in depth understanding of the subject.

What is needed are books that travel a road somewhere between the boundaries outlined above. The books should provide a complete treatment of the subject matter but limit mathematics to a moderate level such as high school algebra and basic statistics.

Wind Turbine Technology is such a book. Wind technology cannot be understood without some mathematical development of the concepts. However, a firm grasp of the subject is possible without delving into calculus and matrix algebra. This book is an attempt to achieve this goal. Where possible, concepts will be explained and developed using simple mathematics. While a full derivation of mathematical principles of some subjects may not be possible, the author believes that a reasonably in depth understanding of the subject matter can still be obtained by explanation and the use of relatively simple mathematics such as basic algebra. Wherever possible, the book relates new concepts to those encountered in everyday life by analogy, helping the reader to more easily understand the explanations.

It is hoped that a book at this level will appeal to a large audience. Wind turbines are rapidly becoming a part of our landscape. Wind turbines are here to stay. The technology will be subject to rapid advances. This will create new jobs for scientists, engineers, manufacturers, installers, maintenance crews, teachers, administrators, financiers and many others in the supply chain. Many of the people in this new industry may already have engineering or science backgrounds and have already graduated from a two or four year institution. This book will hopefully be a perfect book for these people because it would allow them to quickly and efficiently acquire an in depth understanding of the subject.

Many colleges and universities are now offering courses in wind technology and more are being added every year. This book might serve as an ideal text for a one semester course in a two or four year college. Because of the limited math requirement of a course based on the book, it would be open to a much wider enrollment than a purely theoretical course.

Our future and the future of the world depends on adopting and advancing renewable energy. We need to encourage students and others to learn about the technology. We need to find ways to make it easier for people to acquire this knowledge. The author hopes this book will be a small step in that direction and that the book helps to make wind technology more accessible to a world that will increasingly depend on it.

The author's approach is to provide a series of economically priced books in both paperback and as eBooks, each covering one of the several areas of subject matter necessary for a person to have a complete understanding of how wind turbines work. By making several books available, readers who may have interests limited to a specific area of knowledge need not expend a substantial sum of money for information that they may never need or want. As of this writing, the author is preparing additional volumes in this series on wind power technology. A few of the upcoming books will cover subjects such as available energy in the wind, electricity and electronics applicable to wind turbines, wind turbine mechanics and drive trains, hydraulics, generators, control systems and other topics relating to wind turbine design. The author welcomes your feedback on this book.

Contents

About the Author 3
Other Books by the Author 4
Preface 5
Introduction 11
Wind Force 12
    Introduction 12
    Wind Force Explanation 12
Airfoils 18
    Leading Edge 19
    Trailing Edge 19
    Airfoil Upper and Lower Surfaces 20
    Chord Line 20
    Camber Line 22
    Angle of Attack 22
    Relative Wind 22
    Pitch Angle 26
How Airfoils Produce Lift 28
    Introduction 28
    Lift and Drag 28
    Bernoulli's Principle 28
    Momentum Theory of Lift. 34
    Reynolds Numbers 34
    Viscosity 35
    Boundary Layer 37
    Laminar Flow 39

    Airfoil Lift and Drag 42

    Lift and Drag Coefficients 44

Airfoil Development 50

    Introduction 50

    Different Airfoil Requirements for Different Wind Turbine Types 50

    Airfoils Designed for Wind Turbines 53

Power Production of a Wind Turbine Rotor 58

    Airflow in the Vicinity of the Rotor 58

    The Betz Limit 60

    Wake Rotation 61

    Calculating the Power in the Wind 62

    Blade Element Momentum Theory 63

    Blade Tip Losses 66

    Blade Tip Speed Ratio 67

    Solidity 69

    Number of Blades 70

    Stall Regulated Wind Turbines 71

Blade Analysis Example 78

    Introduction 78

    Design Considerations 78

    The Example Design 80

    Calculating the Chord for Each Station 84

    Calculating the Blade Twist 87

    Preparing the Data for Input into WT_Perf 89

Appendix 97

INDEX 99

Notes

Introduction

Wind turbines produce electrical or mechanical power by extracting energy from the wind. Although many early wind turbines relied upon drag forces to produce power, modern turbines rely upon aerodynamic lift forces. Lift forces are generated by airfoils. In this book we will explore the aerodynamic principles that govern how and why airfoils are able to produce lift. We will discuss lift and drag and how wind turbine rotors extract energy from the wind. We will look at airfoil types and some of the characteristics that affect their suitability for use on wind turbine rotors. Additionally, we will learn about the effects of blade speed, numbers of blades and blade shape. We will see how power production, the physical limits on wind turbine efficiency and maximum power extraction from the wind are determined.

Wind Force

Introduction

Wind turbines make use of the force of the wind to generate electricity. The wind force is dependent on wind speed, air density and the area of the object that is exposed to the wind. One of the most important aspects of the wind force is the load that the wind places on the wind turbine, the turbine rotor and the structure (tower) that supports the turbine. Design engineers need to understand these loads in order to successfully design the wind turbine components.

Wind Force Explanation

We all know that standing outside on a very windy day, we will feel a force from the wind. Think of the term "Force" as something you would feel if a person standing in front of you placed both their hands on your shoulders and slowly pushed you away from them. The "push" that you feel is a force. If you are standing outside facing into the wind and there is a very strong wind blowing, your body will feel a force pushing you backwards just like you felt when the person standing in front of you was pushing on your shoulders.

The higher the wind speed, the greater the force. The larger the surface area that the wind can act on, the greater the force. In our example where you are outside facing into a strong wind you would have felt a force pushing you backwards. If you then opened a large umbrella you would suddenly feel a force trying to tear the umbrella out of your hand. If the wind were strong enough you may not be able to stop the umbrella from blowing away. If you were strong enough to hang onto the umbrella the force on the umbrella might be powerful enough to blow both you and the umbrella away!

In order to better understand the wind force, let us begin by considering the density of the air.

Density is defined as the mass per unit volume. Higher density exists when a certain volume has more mass. Lower density exists when a certain volume has less mass. This may be more readily understood by an example. Consider a block of wood in the shape of a cube. A cube means that the dimensions of all six sides are exactly the same. Imagine that you have a 12" cube of wood – a "cubic foot". Let's say it weighs 30 pounds. You probably would not have much trouble lifting it off the ground. Now imagine that the same size cube was made of lead. It would weigh just over 700 pounds. Unless you were prodigiously strong you would not be able to lift this cube of lead. This weight would be at or above the upper limit for the weight-lifting bench press world record. We see from this example, and it is probably quite intuitive that, considering things of the same size, the weight of something is dependent on what it is made from. Lead weighs much more than wood. We can therefore say that lead is more dense than wood. In our example, lead is 23 times heavier than our block of wood. Comparing wood to air, obviously, wood weighs much more than air. In fact, air only weighs about 0.08 pounds per cubic foot, so our example block of wood is 375 times heavier than the same volume of air.

By using the word "density" we are able to describe the mass of something per unit volume. In the case of our wood block we can say that the density is 30 pounds per cubic foot. The density of lead is 700 pounds per cubic foot. The density of air is 0.08 pounds per cubic foot.

So, why does the density of air have an effect on the force it exerts on an object? Well, suppose you are napping at the beach lying on your stomach on your beach blanket soaking up the sun and a friend places a 1 cubic foot block of Styrofoam (weighing only 2 pounds per cubic foot) on your back. This isn't much weight so you might not even notice it. Now, suppose that instead of Styrofoam your friend places a 30 pound wooden cube on your back. You would certainly notice that! We won't even touch upon the effect on your back caused by a cubic foot of lead! So, we can see that the density of an object has a substantial effect on the force which it is capable of exerting. In the case of the block placed upon your back the force was caused by the earth's gravity. In the case of the force caused by the wind, it results from a moving mass of air.

In this section we are not really interested in the mass of wood or lead, except for its use as an example to help us understand the concept of mass. We are primarily interested in the density of air, because that's what wind turbines use to generate power. Air isn't very dense compared to most materials that we are familiar with. As we have already noted, the density of air is about 0.08 pounds per cubic foot.

The equation for density (ρ) is:

$$\rho = \frac{m}{V} \tag{1}$$

where m is the mass and V is the volume. The symbol ρ is the Greek letter Rho.

Air density varies with temperature, pressure and humidity. At sea level and at a temperature of 59°F (15°C) air density is 0.0765 pounds per cubic foot (1.225 kg/m^3). Air density is also measured in slugs per cubic foot. At 15°C, air density is 0.0023769 slugs/ft^3. We will use slugs per cubic foot in our calculation examples.

We have seen that density will affect force so let us consider the force of a moving fluid such as air, which we call "wind". The force exerted by the wind can be calculated using the following equation:

$$F = A * P \tag{2}$$

where F is the wind force, A is the area which the wind is acting upon and P is the pressure exerted by the wind. Pressure is the force per unit area. Pressure in the United States is usually measured as pounds per square inch or pounds per square foot.

If we have a 30 pound weight and apply this weight to an area 1 inch square the force is 30 pounds per square inch (PSI). If we have a 30 pound weight and apply to an area of one square foot we have a force of 30 pounds per square foot (PSF). Note that a force of 30 PSI is 144 times greater than a force of 30 PSF. This is because there are 144 square inches in one square foot. So, when you apply 30 pounds uniformly to an area of one square foot (144 square inches) the force per square inch is $30/144 = 0.21$ PSI.

One example of pressure familiar to most people is the air pressure in the tires of their car. This is usually measured in pounds per square inch (PSI). Most modern passenger cars recommend a tire pressure of 30 to 35 PSI. PSI is handy when dealing with small things like tires which do not have much area. When dealing with large things like the swept area of a wind turbine which may be hundreds of square feet it is more convenient to use pounds per square foot (PSF).

We can calculate the pressure exerted by a moving stream of air using the equation:

$$P = \frac{1}{2}\rho U^2 \tag{3}$$

where ρ is the air density and U is the wind velocity.

When designing wind turbines, we are usually more interested in the wind force on the turbine than the air pressure. We can solve for this force by combining equations (2) and (3), which essentially multiplies the pressure times the area upon which the pressure is acting. Doing so we have:

$$F = \tfrac{1}{2}\rho U^2 A \qquad (4)$$

It is important to notice from equation (4) that although the wind force varies linearly with air density and area, it varies exponentially with wind velocity.

For those who may not commonly work with exponents it is worth taking a moment to explain how exponents work. Notice that the U term is followed by a superscript 2. The "2" is an exponent. This means that the value of U is multiplied by itself, i.e., it is "squared". A number squared is simply that number multiplied by itself. For example, the number 2 squared is written as 2^2. If we square 2 the result is 4 (2*2=4). When we deal with larger number the result can become very large very quickly. For example, 10 squared is 100 (10*10=100). Comparing our two examples, although 4 is only twice as large as 2, 100 is 10 times larger than 10.

Because the wind velocity "U" is squared, **small increases in wind velocity result in large increases in force**. This fact is an extremely important thing to know when designing wind turbines.

The remaining terms in equation (4) are air density, ρ and the area, A. These terms are linear so if either of these variables change, the resulting force only changes in proportion to the change in the variables. The result does not change exponentially. For example, if the swept area of the turbine rotor (A) is doubled, the force on the rotor is doubled. Contrast this to the wind velocity. If the wind velocity doubles, the force increases by a factor of 4.

Now that we have an understanding of the concept of wind force, let us look at a couple of numerical examples showing how the actual calculations are made.

Assume that you are standing outside facing into the wind and holding up a sheet plywood which measures 4 feet by 4 feet (16 square feet). Assume US Customary units. Assume that the wind is blowing at 20 MPH. Using equation (4) we have:

$$F = \frac{1}{2}0.00238 * 20^2 * 16 = 7.6 \, pounds$$

The calculated force on the plywood is just under eight pounds. Let now us double the wind speed to 40 MPH.

$$F = \frac{1}{2}0.00238 * 40^2 * 16 = 30.5 \, pounds$$

You can see that by doubling the wind speed we have increased the force by a factor of four. If we were to increase the wind speed to hurricane speed (74 MPH), the force on the plywood would be in excess of 100 pounds. The exponential relationship between wind force and wind speed explains why high winds can be so damaging to trees, buildings and other structures.

Now that we have a basic understanding of wind force let us take a look at airfoils. Airfoils are shapes that airplane wings and wind turbine rotors can use to efficiently extract energy from the wind.

Airfoils

An airfoil is an aerodynamic shape. It is the cross-sectional shape of an airplane wing, propeller blade, wind turbine blade, bird's wing, sailboat sail or other device designed to produce aerodynamic lift. Some examples of airfoil shapes are shown in Figure 1.

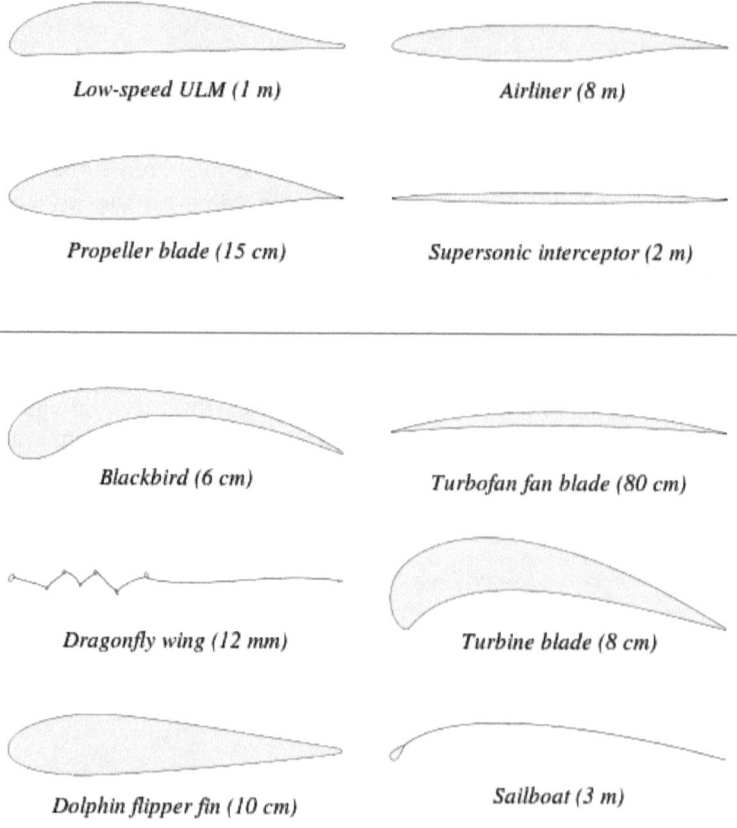

Figure 1 - Airfoil Examples

There are a number of terms which describe certain characteristics of airfoils. The reader is advised to become familiar with these terms as we will refer to them throughout this text. The terms are shown in Figure 2.

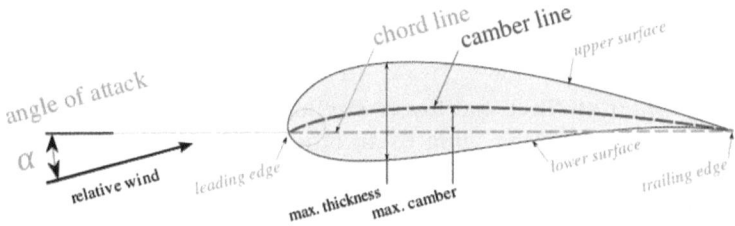

Figure 2 Airfoil Nomenclature (Courtesy Wikipedia)

Leading Edge.

If the airfoil shown in Figure 2 were a section of an airplane wing, the leading edge would be the first point on the airfoil that contacts the wind when the airplane is moving through the air.

During normal operation of a wind turbine blade, the **leading edge** is the first point on an airfoil that contacts the wind as the rotor turns. It can also be defined as the point which would yield the maximum chord length.

Trailing Edge.

The **trailing edge** is the last point on an airfoil that is in contact with the wind. It can also be defined as a point opposite the leading edge which would give the maximum chord length. The leading edge is usually rounded (shown as the small circle in Figure 2) and the curve is sometimes called the leading edge radius. The trailing edge is usually made quite sharp so as to have the least adverse effect (turbulence creation) on the departing air stream. For airfoils operating at the relatively high speeds of modern wind turbine rotors sharp trailing edges are necessary in order to keep acoustic emissions (noise) to a reasonable level.

Airfoil Upper and Lower Surfaces.

The **upper surface** and **lower surface** are self-explanatory when used to describe the wing of an airplane - at least an airplane in normal flight. When the airfoil is used on the rotor of a wind turbine, the lower surface faces the direction from which the wind is blowing, so in Figure 2, the wind would be coming from the bottom of the page.

Chord Line.

The **chord line** is a straight line extending from the leading edge to the trailing edge. A common airfoil, called the Clark Y Airfoil, is shown in Figure 3. The chord line is shown in this figure as the horizontal line labeled on the vertical axis as zero.

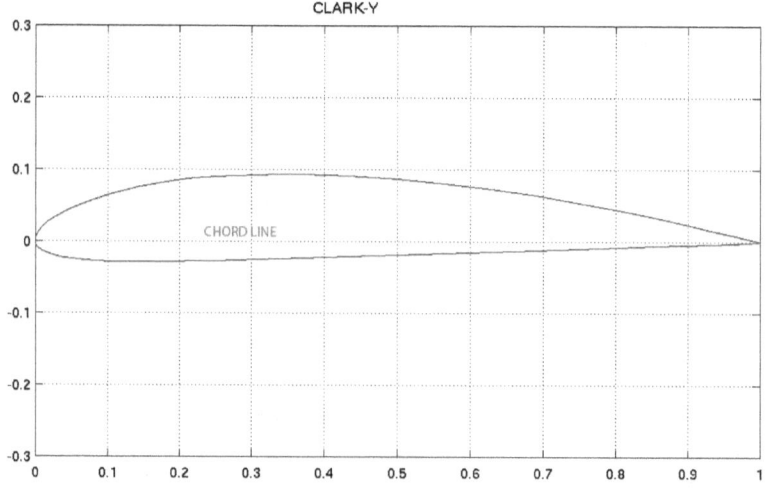

Figure 3 - Clark Y Airfoil

When drawing the shape of an airfoil the chord line is used as a baseline for offsets to the airfoil upper and lower surfaces. This means that all that you need to draw the shape of an airfoil is a grid similar to the one shown in Figure 3 and a table of stations and offsets. The table shown in Figure 4 provides the offsets that we need to draw the pictured airfoil. The top row gives stations along the horizontal axis of the chord line. The chord line is labeled as decimals beginning with 0 and ending with 1. This is a very convenient way to show the stations because we can use the decimals to draw an airfoil of any size simply by multiplying the decimal station by the chord length of our actual airfoil. For example, suppose we wanted to draw an airfoil with a 10 inch chord. Consider the 3rd column which is the chord station 0.2. Notice in Figure 3 that the vertical lines are labeled as decimal fractions of 1. To calculate the 0.2 station of our airfoil with the 10" chord we can multiply 10 inches by 0.2 to calculate the actual distance along the chord from the zero station. The actual station would be 10" * 0.2 = 2".

Station	0	0.1	0.2	0.3	0.4	0.5	0.6	0.7	0.8	0.9	1
Upper	0	.063	.084	.090	.091	.086	.076	.061	.044	.023	0
Lower-	0	.015	.030	.026	.023	.019	.015	.012	.08	.004	0

Figure 4 Clary Y Airfoil Stations and Offsets (approximate)

The offsets are measured at 90° to the chord line. The table gives the distances to the upper and lower airfoil surfaces for each station. (These offsets are for demonstration purposes only. Do not try to use them to create an actual airfoil.) The offsets to the upper surface have a positive sign and those to the lower surface have a negative sign (there wasn't room in our table to insert a minus sign before each offset so it was omitted). In order to draw an actual airfoil of a particular size we will need to perform the same calculation as we did for the stations. For example, the upper offset at station 0.2 is 0.084. Using our example airfoil with the 10 inch chord the distance from the baseline to the upper surface would be calculated as: 0.084 * 10" = 0.84". We would use the same procedure at each station to draw the lower airfoil surface.

It is worth noting that since the advent of the airplane, hundreds of airfoils have been designed. Offsets for many of these airfoils can be found online and in publications.

Camber Line.

The **camber line**, sometimes called the mean camber line, is a curved line halfway between the upper and lower surfaces. The **maximum camber** is the largest dimension between the chord and the camber line. The **maximum thickness** is the maximum distance between the upper and lower surfaces.

Angle of Attack.

The **angle of attack** is the angle between the chord line and the relative wind. In aerodynamics the angle of attack is usually designated by the Greek letter alpha (α). The angle of attack of an airfoil is very important because it determines the amount of lift that the airfoil will develop, and ultimately, the power that a wind turbine will produce. The amount of lift is directly related to the power production of a wind turbine rotor. We will discuss these concepts in detail.

Relative Wind.

Because the **relative wind** is such an important concept, a student of wind turbine aerodynamics must thoroughly understand it. Sometimes the term apparent wind is used instead or relative wind but, in our context, they mean the same thing. An example of relative wind should help make the concept clear. The tip of a rotating wind turbine blade is shown in Figure 5. To better understand the blade orientation shown in this figure just image that you are standing on the ground at the base of a tower looking straight up at a wind turbine just at the moment when one of the blades passes directly overhead. At that moment you would be looking at the end of the blade (blade tip) as it passes by. Figure 5 shows what the blade tip would look like if you were to snap an image of it with your camera at exactly the instant it passes overhead.

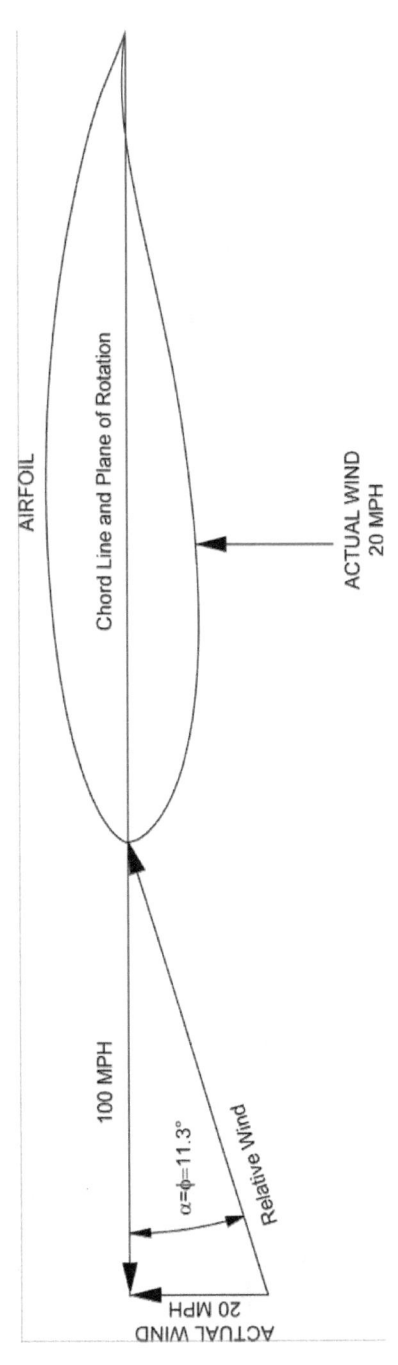

Figure 5 - Relative Wind

The actual wind direction and velocity is shown as the vector in the figure labeled *Actual Wind*. This is shown in two places in our figure. If you are unfamiliar with vectors, the concept is quite simple. You can think of a vector as an arrow of a certain length that points in a certain direction. A vector describes both magnitude and direction. When a vector is used to describe the wind, the length of a vector indicates the speed of the wind. The direction that the vector is pointing tells us the direction from which the wind is blowing.

It is important to understand that in our example in Figure 5 (and in the case of an actual wind turbine) the speed of the turbine blade is considerably faster than the actual speed of the wind. In our example, the tip of the blade is travelling at 100 MPH but the actual wind speed is only 20 MPH. So, the blade tip is travelling 5 times faster than the wind. The lengths of the vectors provide the reader with a graphic representation of the velocities. Although the drawing is not drawn to scale, you can see that the vector labeled "100 MPH" is longer than the Actual Wind vector of 20 MPH.

The direction of the actual wind is 90° to the plane of rotation of the turbine blade. The vector labeled "100 MPH" shows the direction and speed of the blade tip in vector form. In the image we copied the Actual Wind vector and located it exactly at the end of the blade speed vector (100 MPH vector). We can now draw a new vector to complete the triangle. We label this vector "Relative Wind". You can see the Relative Wind vector in Figure 5. You can also see that the Relative Wind vector makes an angle of 11.3° with the plane of rotation of the blade. This angle is the Angle of Attack and it is designated by the symbol α (alpha).

One way to think about relative wind is to imagine that you are sitting on the end of the blade as it is moving through the air. If there were no actual wind you would feel the wind coming from exactly the direction that you were moving, i.e., straight ahead. If the actual wind began to blow from the direction shown in

Figure 5 you would feel a shift in the wind so that the wind that you actually felt would be coming from the direction of the Relative Wind vector. In other words, the wind would shift slightly to your left. This is not much different than riding on your bicycle on a day with no wind. You would feel the wind coming from the exact direction in which you were moving. If a wind began to blow from a direction 90° to the direction of travel you would feel a shift in the wind toward the direction from which the wind was blowing. Assuming that your bicycle does not change speed, the amount of shift (the angle of attack) would depend on the velocity of the actual wind. The greater the velocity, the greater the angle of attack. You may want to draw your own set of vectors with a blade speed of 100 MPH and a wind speed of 100 MPH to prove to yourself that the angle of attack would be 45°.

Expressed in mathematical terms, it can be seen from Figure 5 that the angle and magnitude of the relative wind is the vector sum of the blade speed vector and actual wind vector. The angle between the plane of rotation and the relative wind is given by:

$$\varphi = \tan^{-1}\left(\frac{Actual\ Wind\ Speed}{Blade\ Speed}\right) \qquad (5)$$

The function $\tan^{-1}$ is the arc tangent of the angle.

In our example shown in Figure 5 the actual wind speed is 20 MPH and the blade speed is 100 MPH, so we can calculate the relative wind angle as follows:

$$\varphi = \tan^{-1}\left(\frac{20}{100}\right) = 11.3°$$

Dividing 20 by 100 gives 0.200. The arc tangent of 0.200 is 11.3 degrees. It is worth noting that the angle of attack is also 11.3° because, in this example, the chord line and plane of rotation are collinear. In actual turbine rotor designs however, the chord line may not be exactly coincident with the plane of rotation. As we shall see next, in an actual wind turbine rotor there may be a small angle between the chord line and plane of rotation.

Pitch Angle.

In the Figure 5 the chord line lies on the plane of rotation so the angle between the chord line and plane of rotation is zero. However, this is usually not the case. In an actual wind turbine rotor there will usually be a small angle between the chord line and plane of rotation. This angle is called the pitch angle. The pitch angle could be a positive angle or a negative angle.

Consider Figure 6 which shows a blade tip with a pitch angle of 3.3°. The blade tip is pitched in a positive direction (toward the actual wind direction) by 3.3°. The angle between the relative wind and plane of rotation is the same as in the previous example (11.3°). However, the angle of attack is now 8° (11.3° - 3.3° = 8°).

Remember, the angle of attack is defined as the angle between the chord line and the relative wind. It is not the angle between the plane of rotation and the relative wind.

Now that we have an understanding of some of the terminology relating to airfoils and their geometry relative to the wind we will next consider how airfoils actually produce lift.

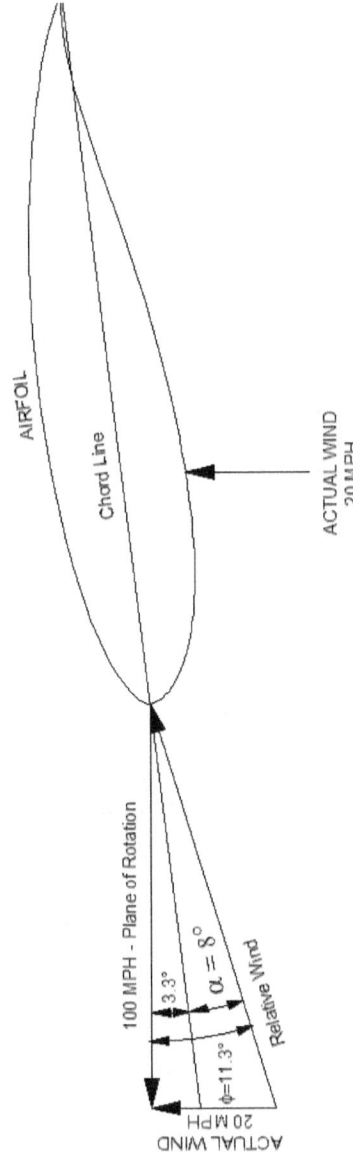

Figure 6 – Pitch Angle.   Chord not Parallel to Plane of Rotation

How Airfoils Produce Lift

Introduction

A wind turbine blade is able to extract power from the wind because the shape of the blade airfoil moving through the airstream produces a lift force. There are a number of explanations of exactly how this happens. The first explanation that we will consider relies upon differences in air pressure in the vicinity of the upper and lower surfaces of the airfoil. This explanation is referred to as Bernoulli's principle.

Lift and Drag.

When aerodynamic forces act on an airfoil in motion these forces create lift and drag. The lift created when an airplane wing moves through the air is what allows the airplane to overcome gravity and lift off from the ground. As nothing in this world is really free, the aerodynamics which produce lift (a beneficial force if you want to leave the ground) also creates drag. Drag acts against the direction of motion and in an airplane, this drag must be overcome by the engine and propeller. In this section we will explore this lift force beginning with Bernoulli's Principle.

Bernoulli's Principle

Daniel Bernoulli was a Swiss scientist who published this principle in 1738. The theory states that when the velocity of a fluid increases there is a decrease in pressure. Although most people think of water as a fluid it is important to understand that in our context air is also considered a fluid. Bernoulli's theory can be visualized by looking at Figure 7 which shows a stationary cylinder in an airstream.

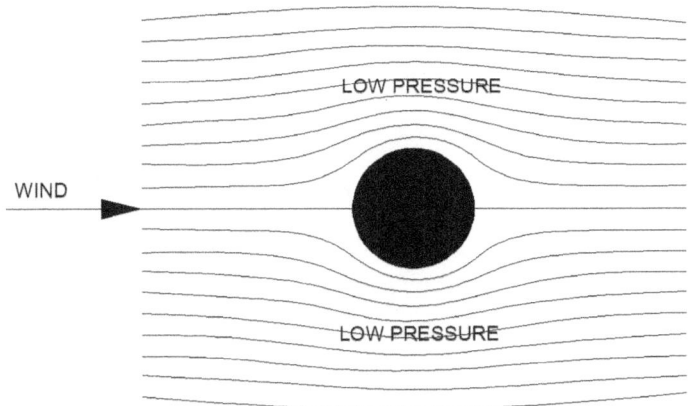

Figure 7 - Flow Around a Cylinder

Imagine that the stationary cylinder in our figure is solid column such as a tall cylindrical chimney located in an open field on a windy day. You are located at the top of the chimney looking down toward the ground. The horizontal lines are **streamlines** which represent paths taken by individual particles of air. As the wind approaches the chimney the streamlines begin to curve around the structure. The nearer the chimney, the more the streamlines have to curve in order to flow around the chimney. Notice that far upstream and far downstream of the chimney (to the left and right in our image) the wind flow streamlines are equally spaced. However, notice how they become crowded together as they pass around the chimney.

In order for all of the air particles to arrive at the right edge of our image at the same instant, it is necessary for the streamlines to speed up as they pass by the chimney. They have to speed up because the curved path is a longer path than the nearby straight paths. Because the air near the chimney has further to go than the air away from the chimney it must travel faster to arrive at its destination at the same time as the air further away from the chimney. At some distance downstream of the chimney, the streamlines resume their original velocity. **Because the air velocity increases as it flows around the chimney, the air pressure decreases** reaching its lowest value at a point halfway around the chimney. This area is labeled "Low Pressure" in our image.

If you have an object immersed in an airflow – imagine an airplane wing – and the air pressure decreases on one side of the object, the pressure differential between the two sides will cause the object to move away from the high pressure area toward the low pressure area. In aerodynamics this effect is referred to as lift.

In Figure 7 the chimney is a perfect cylinder so there are two identical low pressure areas, one on each side of the chimney. Because of this symmetry, there is no net lift produced. The low pressure on one side is exactly equal to the low pressure on the other side.

Let us now consider what happens when we place a rotating cylinder in a moving air stream as in Figure 8. This figure illustrates that a rotating cylinder causes the air flow to speed up on one side of the cylinder and slow down on the other side. The velocity of the air is increased on the side of the cylinder that is moving in the same direction as the air flow and decreased on the side of the cylinder that is moving against the airflow. This effect is caused by friction and drag between the surface of the cylinder and the moving air. One way to think of the effect is to imagine that the spinning cylinder surface helps to push the air along on the side where it is moving in the same direction as the wind and it retards the wind on the side where it is moving against the wind.

Unlike the stationary cylinder, the rotating cylinder causes asymmetry in the airflow. There is negative pressure on the side where the air streamlines have increased in velocity and positive pressure on the side where the air streamlines have slowed. The result is a pressure differential between the two sides. The rotating cylinder creates a lift force directed 90° from the direction of the wind.

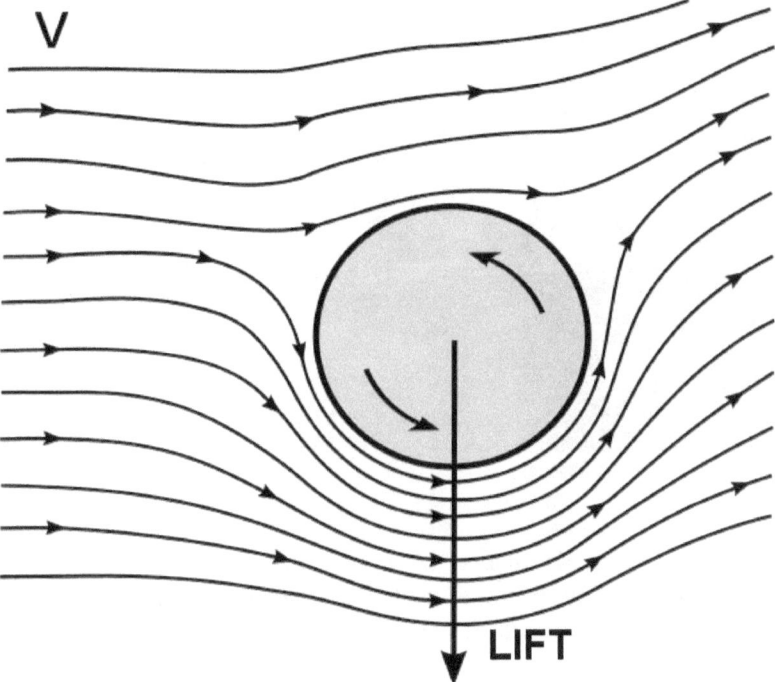

Figure 8 - Flow around a Rotating Cylinder

The lift produced by a cylinder rotating in a moving fluid is referred to as the Magnus Effect and the cylinder is often referred to as a Flettner Rotor after Anton Flettner who was the first person to use this effect for ship propulsion in 1924. See Figure 9 which shows the ship with the two tall rotating cylinders. Because the Flettner Rotor requires power to spin the cylinder it may not be the most practical solution for a wind turbine blade. Fortunately, a similar result can be produced by an airfoil.

Figure 9 - Flettner Rotors.

When an airfoil is placed in a moving stream of air and the airfoil is correctly oriented, it is capable of producing lift. This is illustrated by means of a typical airplane wing in Figure 10. As the moving air passes over the upper surface of the airfoil it is accelerated causing a drop in pressure relative to the lower surface and relative to the air pressure far upstream. The pressure differential between the upper and lower surfaces of the airfoil produces lift.

The pressure distribution along a typical airfoil is shown in Figure 11. The vectors above the airfoil indicate negative pressure and those below, positive pressure. A large low-pressure area exists at a position centered at about the 1/4 chord position (1/4 of the way from the leading to trailing edges of the airfoil). The result of the pressure distribution is a lift force in the vertical direction.

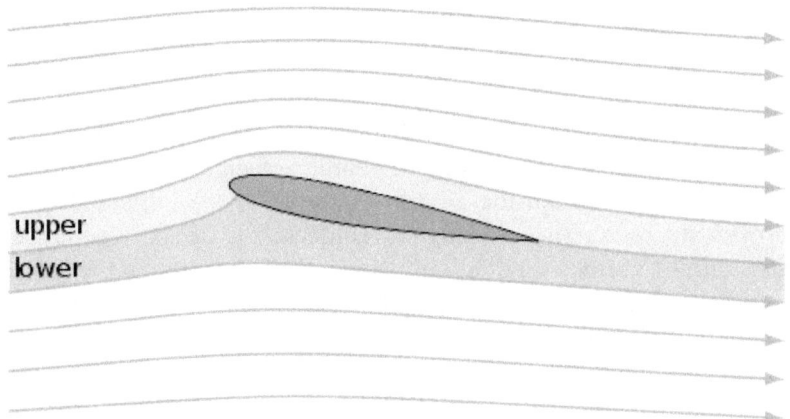

Figure 10 - Airfoil in a Moving Airstream

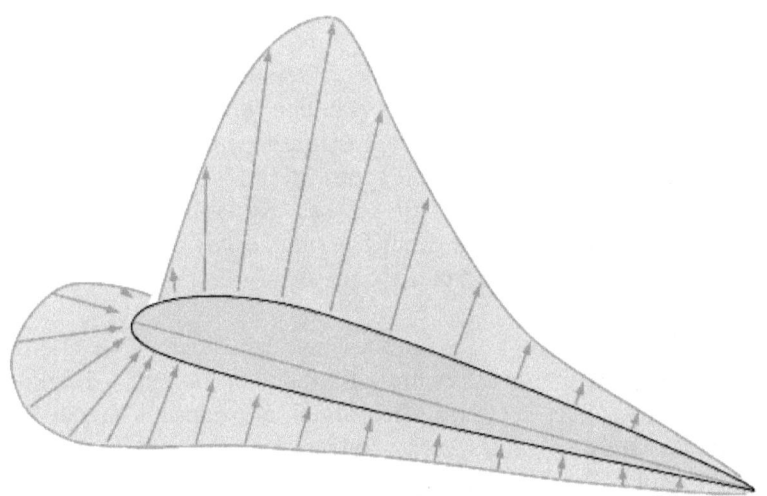

Figure 11 - Airfoil Pressure Distribution.

Momentum Theory of Lift.

Lift can also be explained in terms of Newton's laws. According to Newton's second law, the resulting force on an object is equal to its rate of momentum change. An airplane wing passing through the air causes the air to be deflected in a downward direction. This is visible in Figure 10 where the streamlines are deflected slightly toward the bottom of the page by the airfoil. Newton's third law states that for every action there is an equal and opposite reaction. In the case of the airplane wing, as the wing deflects air downward, the air exerts an equal upward force on the wing. So, by forcing air downward the wing is forced upward. The same principle can be applied to a wind turbine blade. The blade deflects the airflow which in turn causes the rotor to turn.

Reynolds Numbers

In order to develop an in depth understanding of how wind turbine blades can efficiently produce energy, we need to become familiar with the concept of the Reynolds number (Re). Reynolds numbers apply to all types of fluid dynamics, such as the flow of water, oil or air. The Reynolds number is a dimensionless number that describes the ratio of inertial forces to viscous forces of the fluid under consideration.

In aerodynamics, Reynolds numbers are useful when we attempt to categorize whether flows are laminar or turbulent. Laminar flow occurs when the fluid streamlines are parallel and do not intersect with each other. Turbulent flow, on the other hand, occurs when fluid particles are moving in random directions. The flow is chaotic. Anyone who has stood on a dock in a moving river and seen the eddy currents around and downstream of dock pilings has observed turbulent flow. Because of the random and chaotic movement of the particles turbulent flow tends to be more wasteful of energy than laminar flow.

In general, laminar flow occurs at low Reynolds numbers and turbulent flow occurs at high Reynolds numbers.

Reynolds numbers are also useful in predicting the performance of different sized airfoils. There is a substantial difference between the Reynolds number of a small wind turbine having a chord measured in inches and the Reynolds number of a large utility scale turbine having a chord measured in feet. Reynolds numbers have proven useful in designing specific airfoils for wind turbine blades of different sizes.

A Reynolds number for a wind turbine blade can be calculated using the following equation:

$$R_c = \frac{UL}{v} \tag{6}$$

where U is the velocity of the relative wind, L is the length of the airfoil, usually measured along the chord, and L is the kinematic viscosity of air. Kinematic viscosity is calculated from:

$$v = \frac{\mu}{\rho} \tag{7}$$

where μ is dynamic viscosity (or just viscosity) and ρ is the air density.

Viscosity

Viscosity is a measure of a fluid's resistance to gradual deformation by shear stress or tensile stress. Imagine that a very thick and sticky substance such as cold molasses is poured onto a flat surface such as a kitchen counter and a flat bottom frying pan is set on top of the molasses – noted as "Boundary Plate 2D Moving" in the image. Assume the molasses is the green fluid in Figure 12 and the boundary plate at the bottom is the kitchen counter. If you attempt to slide the frying pan over the counter, the molasses will resist the motion and it will take considerable effort on your part to overcome the friction. When you slide the pan over the counter you are subjecting the molasses to a shear stress.

Kinematic viscosity v is the viscosity divided by the density of the fluid. For air at 59°F (15°C), v is 0.0001569 Ft²/S (feet squared per second). When discussing viscosity, you may sometimes run across the term "inviscid" flow or inviscid fluid. This simply means that the fluid or flow is ideal in that it has zero viscosity. This is a characteristic that would not be encountered in the real world, but it is sometimes convenient for aerodynamic theory or modeling.

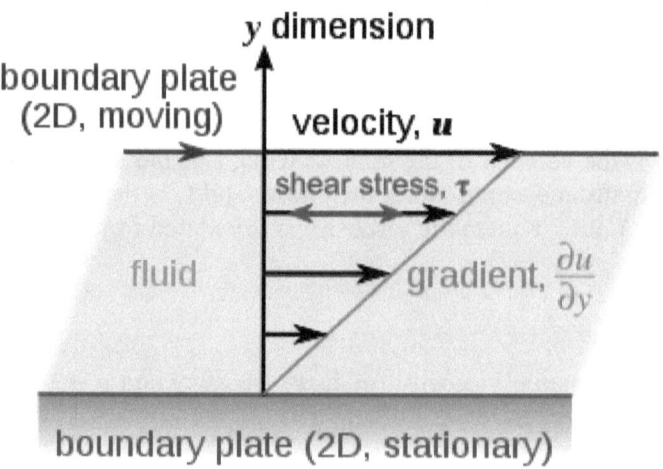

Figure 12 - Viscosity of a Fluid

A shear stress occurs when the fluid particles slide past each other. In our molasses example in Figure 12, as you slide the pan, a thin layer of the molasses near the counter top will remain stationary and attached to the counter while a thin layer attached to the bottom of the pan will move with the pan. The molasses particles between these two surfaces will move a certain distance in the direction of the pan movement dependent upon where a particular particle is located vertically in the column. Particles near the counter will move less than particles near the pan. Because cold molasses has a high resistance to deformation in shear we would say that it has a high viscosity. If, instead of molasses we used cooking oil, it would require much less effort to slide the pan and we would say that the oil had a low viscosity. It may seem that something as thin as air would have a very low viscosity, and it does.

We have learned that the Reynolds number gives a ratio of inertial forces to viscous forces. As we have seen, the viscous forces are dependent on the viscosity and density of the fluid. When air flows over the surface of an airfoil, a very thin sheet of air will "stick" to the airfoil surface and remain stationary. As we move away from the airfoil surface the velocity of the air will increase until we are far enough away that the air is moving at its free stream velocity. This happens because friction between the air and the surface of the airfoil causes the air to "stick" to the surface. At the airfoil surface the air has insufficient inertia to overcome the viscous forces.

Boundary Layer

The area near the surface of an airfoil where the velocity transitions from zero to the approaching the freestream velocity is called the boundary layer. A boundary layer is depicted in Figure 13. It may be helpful to think of the black line at the bottom, which represents a solid surface, as a plateau rising above a flat prairie. The area to the left of the black line drops off precipitously so that the wind vectors represented by u_0 are unaffected by surface drag. This causes the upstream vectors u_o to be the same length. There is no wind shear. Saying there is no wind shear is the same as saying there is no change in velocity with height.

As we move to the right, over the surface of the plateau, the velocity vectors near the surface $u_{(y)}$ have shortened indicating a reduced velocity. As height is increased, the wind velocity increases until reaches the value of u_0. All aerodynamic surfaces subject to air flow such as airplane wings and wind turbine blades will have a boundary layer. Because of this boundary layer, the Reynolds number becomes very important in helping to predict the actual performance of the airfoil.

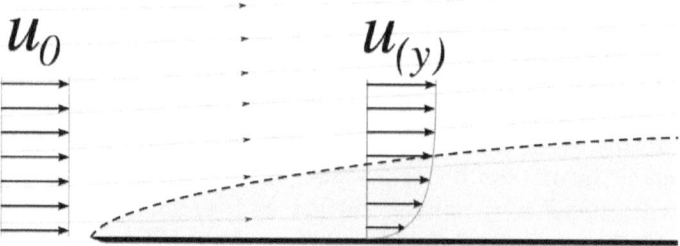

Figure 13 - Boundary Layer

In addition to kinematic viscosity, the Reynolds number contains two other terms. U is the velocity of the air and L is the length of the object that the air is traveling over. For a wind turbine blade the velocity is simply the relative wind speed at the blade radius which is being investigated. Outer portions of the blade travel faster than inner portions. The velocity used in calculating a Reynolds number is not the wind speed, it is the relative wind speed – the relative wind vector shown in Figure 5.

The length term is usually a matter of convention. If we were calculating the Reynolds number of a golf ball the length would be the diameter of the ball. For wind turbine blades the length would be the chord dimension.

Let us look at a couple of examples of Reynolds numbers for typical wind turbine blades. We will compare two turbine blades. One will be a blade for a small wind turbine with a blade tip chord length of 1 foot. The second blade will be for a large turbine with a chord length of 5 feet. We will assume a relative wind speed of 100 MPH (146.66 Ft./Sec.) and a kinematic viscosity of air of 0.0001569 Ft²/S.

For the small blade, using (6) we have:

$$Re = \frac{146.66 * 1}{0.0001569} = 934{,}735$$

For the large blade we have:

$$Re = \frac{146.66 * 5}{0.0001569} = 4{,}673{,}677$$

Notice that because of the linear relation between blade width and Re the large blade has a Reynolds number 5 times the magnitude of the small blade and that the Re magnitudes for both blades are rather large numbers.

Laminar Flow

Laminar flow around an airfoil occurs when the fluid flows in parallel layers or streamlines such as in Figure 14. Notice that even when the layers curve around the airfoil they remain parallel and uniform.

Turbulent flow occurs when the streamlines diverge and become chaotic as in Figure 15. When the flow is turbulent there can be cross currents and circulation between the layers. The flow direction can even reverse and move against the wind direction. We might say that laminar flow is orderly and turbulent flow is disorderly. In laminar flow there is no energy transfer between streamlines.

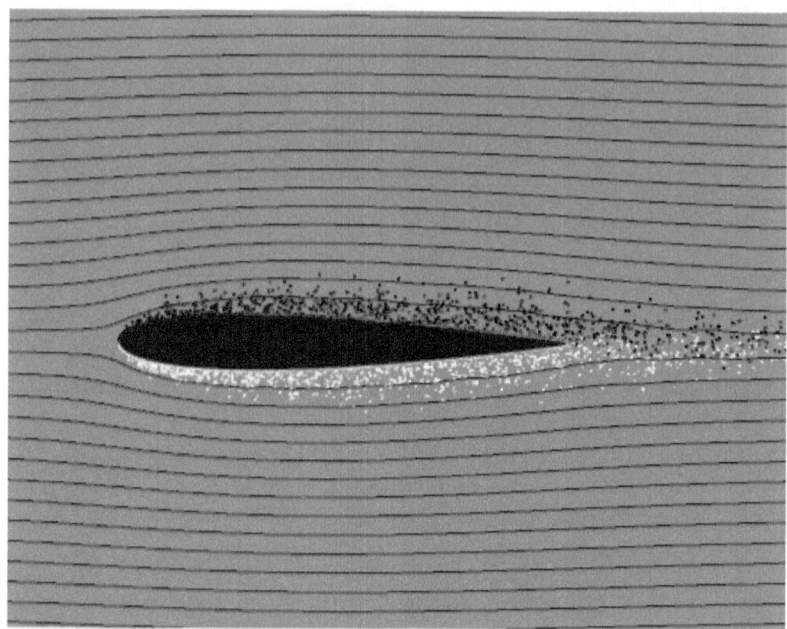

Figure 14 - Laminar Boundary layer

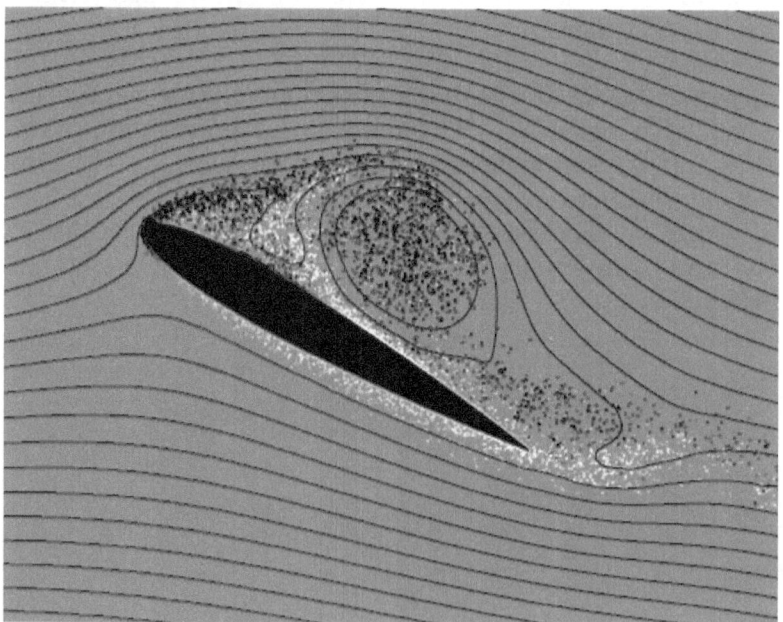

Figure 15 - Turbulent Flow

Laminar flow in a boundary layer attached to a wind turbine blade is shown in Figure 16. The surface of the turbine blade is the hatched area at the bottom of the image. The free stream wind velocity (relative wind speed) is the upper solid line. Notice that at the left side of the image, near the airfoil leading edge, the length of the wind speed vectors do not change as dramatically in length from the airfoil surface to the free stream as they do further along the airfoil. As the air moves along the airfoil the skin friction slows the air particles more and more, increasing the shear and boundary layer thickness.

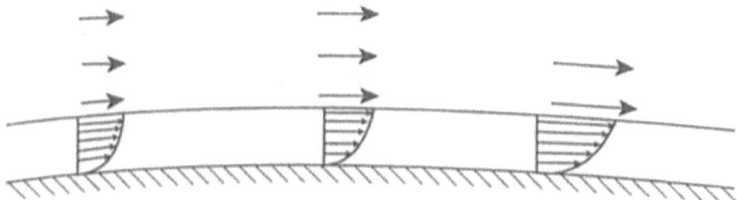

Figure 16 - Laminar Boundary Layer

If the air particles lose too much of their energy, the flow in the boundary layer becomes turbulent as in Figure 17 where the flow detaches and actually reverses direction.

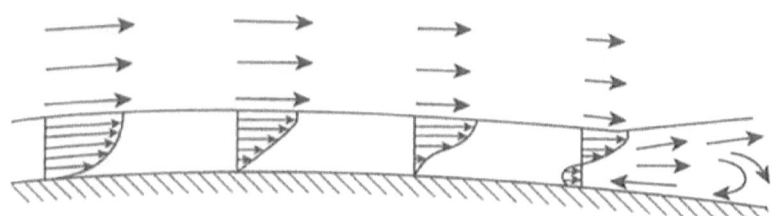

Figure 17 - Turbulent Boundary Layer.

In some cases, laminar flow is desirable because there is less drag than there would be if the flow were turbulent. For small chord lengths (low Re) it is relatively easy to maintain laminar flow because the distance traveled by the airflow is relatively short and skin friction has less of an effect than it would on a large airfoil. However, for laminar flow to exist, the airfoil surface must be very smooth and free from roughness. This is a difficult condition for wind turbine blades because the surfaces are often subject to contamination by bugs and dirt and because they are inaccessible and not easy to keep clean.

Laminar flow can result in adverse pressure gradients between the leading and trailing edges which can cause flow separation as in Figure 17. The air becomes detached from the airfoil surface and it becomes turbulent increasing drag. Although this may seem counterintuitive, flow separation occurs more readily in a laminar boundary layer than in a turbulent boundary layer. One explanation is that in turbulent flow small airflow changes tend to get damped out but in laminar flow a small disturbance can trigger separation. One way to avoid turbulent flow is to delay flow separation by keeping the flow attached to the airfoil for as long as possible by encouraging an early transition to turbulent flow. This is done on golf balls by dimpling the surface or by the fuzzy surface of a tennis ball. On an airfoil, vortex generators are sometimes used.

Large wind turbine blades which have relatively large Reynolds numbers, as compared to small wind turbines, tend to maintain turbulent boundary layers and this results in lower overall drag than airfoils with small Reynolds numbers where the airflow can easily become detached. Special airfoils have been designed by the United States National Renewable Energy Laboratory and others specifically for wind turbines of various sizes allowing them to operate more efficiently at low Reynolds numbers.

Airfoil Lift and Drag

Now that we have an understanding of some of the factors affecting airfoil performance and how airfoils produce lift we can look at the actual forces on an airfoil.

The pressure distributions about an airfoil moving through the air can be resolved into a lift component and a drag component.

This is shown in Figure 18. The vector lengths have been exaggerated for clarity.

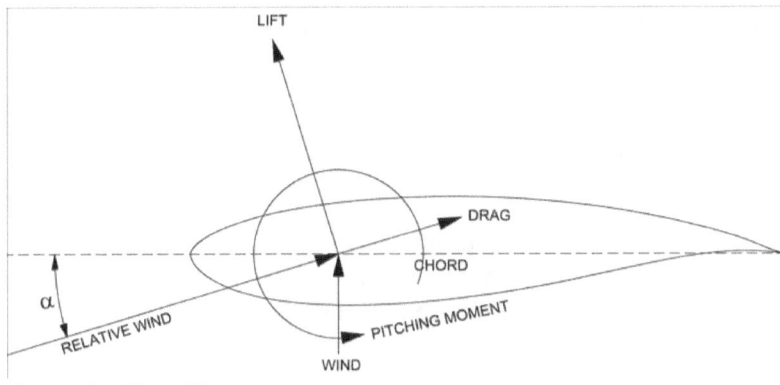

Figure 18 - Lift and Drag

The lift vector is always 90° to the angle of the relative wind. The drag vector is in the same direction as the relative wind (and therefore 90° from the lift vector). We can resolve the lift and drag of an airfoil into thrust as shown in Figure 19.

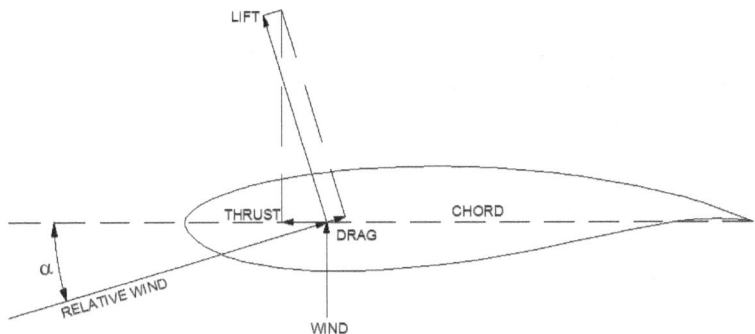

Figure 19 - Airfoil Thrust

The thrust from the airfoil is what provides the torque to turn the rotor shaft of a wind turbine.

It should be pointed out that Figure 18 shows a circular arrow centered on the quarter chord location entitled "Pitching Moment". Aerodynamic forces on the airfoil can cause a twisting force that tends to change the pitch the airfoil. For fixed pitch blades which are rigidly mounted to the hub this force is usually of little consequence. For blades which have some sort of variable pitch adjustment, this force may need to be taken into consideration.

Lift and Drag Coefficients

It is customary to calculate lift and drag using coefficients. A coefficient is simply a number that is used to multiply times a variable. For example, 4X means that the coefficient 4 is to be multiplied by the variable X.

The lift of a section of a wind turbine rotor blade can be calculated using Equation (4) with the addition of a lift coefficient term C_L.

$$Lift = \tfrac{1}{2}\rho U^2 A C_L \tag{8}$$

Similarly, the drag of a section of a wind turbine rotor blade can be calculated using Equation (4) with the addition of a drag coefficient term C_D:

$$Lift = \tfrac{1}{2}\rho U^2 A C_D \tag{9}$$

Both the lift and drag of an airfoil change with the angle of attack α.

A graph of the lift coefficient of a NACA 2415 airfoil is plotted as a function of angle of attack in Figure 20. NACA stands for the National Advisory Committee for Aeronautics. The four digits following the NACA designation are representative of a family of NACA airfoil sections which were developed in the

1930s. These airfoil sections are commonly referred to as the "NACA four-digit" series. Each of the four numbers has a specific meaning which the reader is encouraged to look up. We will note here that the last two digits represent the airfoil thickness as a percentage of chord. For example, the NACA 2415 shown has a thickness of 15% of the chord. There are other 24 series airfoils such as the 2412 and 2418 which have 12% and 18% thickness. A profile of the NACA 2415 is shown in Figure 21.

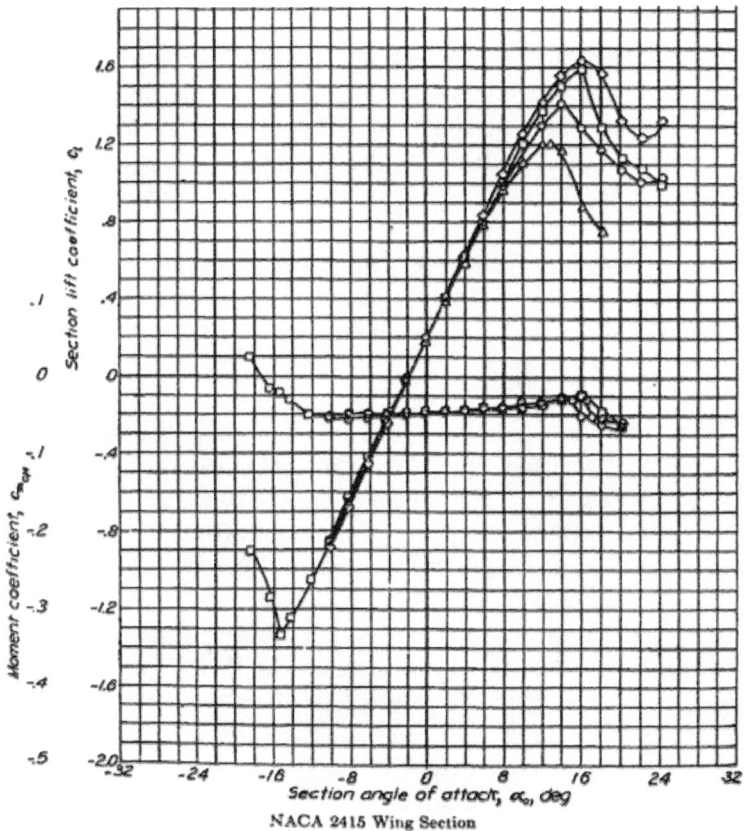

NACA 2415 Wing Section

Figure 20 - Airfoil Lift and Drag

The lift curve shown in Figure 20 is the long, more or less straight line, running from lower left to upper right. The angle of attack of the airfoil is numbered at the bottom horizontal axis

(Section Angle of Attack). The lift coefficients (Section Lift Coefficient) are numbered on the left vertical axis and run from about -1.3 to +1.6. The angles of attack of the airfoil are plotted from about -18° to +26°.

Notice that the figure actually contains four lift curves. Each curve represents a different Reynolds number. For the straight portion of the lift curve the lift coefficients for all of the Reynolds numbers are about the same. Notice however, that at high angles of attack the lift coefficients diverge considerably.

It can be seen from Figure 20 that the airfoil begins to produce positive lift at an angle of attack of -2°. Maximum lift is achieved at an angle of attack of 13° to 16°, depending on the Re. Between these two points the increase in lift with angle of attack is roughly linear.

The graph clearly illustrates the relationship between the maximum lift and Reynolds number. Higher Reynolds numbers allow the airfoil to produce greater values of maximum lift.

Notice from Figure 20 that once the maximum lift coefficient (for a particular Reynolds number) is reached, the lift curve drops precipitously. For example, the lift curve having the lowest peak begins to drop at an angle of attack of about 13°. The lift coefficient drops from about 1.2 at 13° to 0.7 at 18°. The point at which the lift curve suddenly decreases if referred to as "Stall". If this graph were portraying the lift of the wing of an airplane in flight, the stall point would mean that the airfoil could no longer fully support the aircraft in flight – certainly not a positive situation for the pilot passengers in the airplane. Airplane pilots have extensive and recurrent training in procedures required to avoid stall and how to recover from an inadvertent stall. Modern airplanes are designed to be able to recover from stall providing there is enough altitude to do so.

The stall angle is associated with the laminar separation shown in Figure 15. and the detachment of airflow from the airfoil.

The peak of the lift curve is generally referred to as the **stall angle** of the airfoil. It is the angle at which the airfoil begins to produce less lift with increases in the angle of attack. As we shall see in our discussion of stall regulated wind turbines, airfoil stall can be a very useful feature as it effectively limits the power which a wind turbine can produce. As we have already noted with respect to airplane wings, flight in the stall regime is generally to be avoided at all costs as it may result in the aircraft wing's inability to support the weight of the aircraft, sometimes with catastrophic consequences.

Figure 21 - NACA 2415 Airfoil Profile

Airfoil drag coefficients are shown as a function of angle of attack in Figure 22. The graph illustrates that, compared to maximum lift values, drag coefficient values are generally quite small. The graph also shows the typical "bucket" shape of a drag coefficient.

When designing certain aerodynamic devices such as wind turbine blades or glider wings the ratio of lift to drag is very important. A glider remains aloft by extracting the greatest amount of lift from a vertical airstream while minimizing its drag in order to maintain altitude for the longest possible time. For a wind turbine blade we also want to generate the greatest amount of lift for the least amount of drag as this will result in the greatest amount of torque on the rotor shaft for a given wind speed. We can see the lift/drag as a function of angle of attack for an airfoil in Figure 23. Note that as the angle of attack increases so does the efficiency of the airfoil (increasing L/D) until an angle of attach is reached of about 6°. The efficiency of the airfoil then begins to decrease.

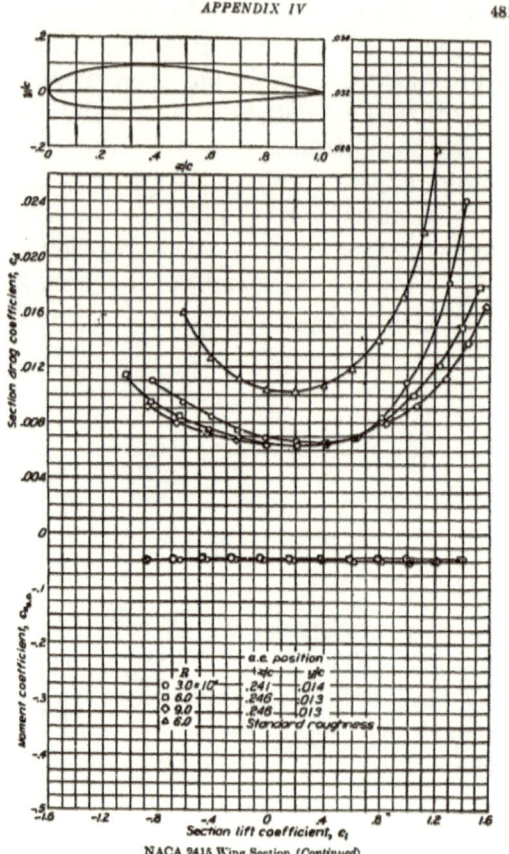

Figure 22 - Airfoil Drag

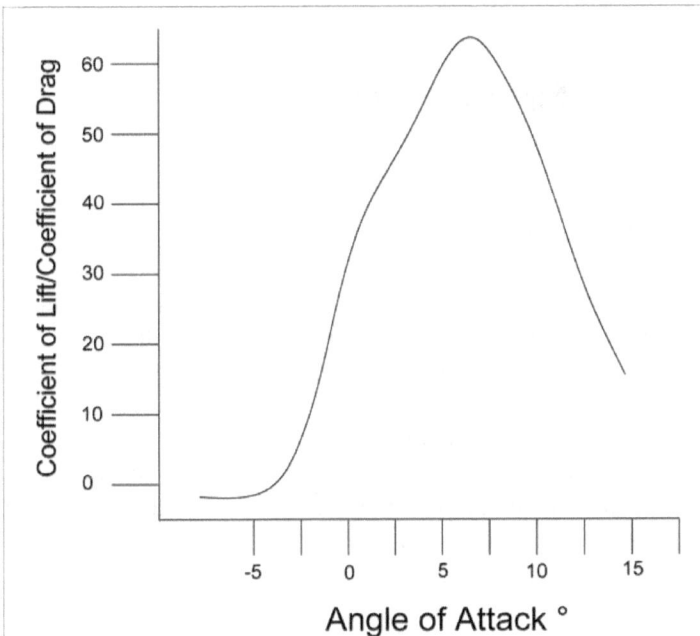
Figure 23 - Lift/Drag

Airfoil Development

Introduction

Until fairly recently, the primary airfoil data that was available for use on wind turbines was data for airfoils designed for aircraft. However, because wind turbines have different requirements than aircraft, the use of aircraft airfoils for wind turbine rotors often resulted in less than optimum wind turbine performance. We have already touched on one of these issues. Aircraft wings can be relatively large compared to some wind turbine blades so wind turbine Reynolds numbers would be lower, and performance would be degraded.

Another issue is that airplanes do not operate in the stall regime. Wind turbine rotors, on the other hand, can spend much of their time operating at high angles of attack. A rough surface on a turbine blade can change its performance. Compared to turbine blades it is much easier to keep airplane wings clean because the turbines are mounted on high towers sometimes hundreds of feet in the air whereas aircraft wings can be cleaned when they are on the ground. These and other issues mean that airfoils developed for aircraft are probably not the best choice for wind turbine rotors.

Different Airfoil Requirements for Different Wind Turbine Types

In a later section we will discuss the two major divisions of horizontal axis wind turbine operation in detail. However, it will be useful for the purposes of this section to have a basic understanding of this topic in order to better understand the different airfoil requirements of these turbine types.

In the first type of horizontal axis wind turbine design the rotor changes its rotational speed in proportion to wind velocity. We call this a Variable Speed Wind Turbine. It turns slowest at low wind speeds and fastest at high wind speeds. In an actual wind turbine design the change in RPM vs. windspeed may not be perfectly linear but for our purposes here we will assume that it is. Because the rotor changes RPM as a function of wind speed, the generator also changes RPM. In this type of turbine, depending on the design, the generator can be connected directly to the rotor. In others, the generator is connected through a gearbox speed-increaser so that the generator turns at some fixed multiple of the rotor speed.

Because the rotor speed increases with wind speed, the angle of attack of the blade remains the same independent of wind speed. This statement is important, so it bears repeating. **The angle of attack of a variable speed rotor is constant and independent of wind speed.** Like most things there are exceptions to the rule but this is a fair statement in general. This should become apparent if you look at Figure 5. If the blade speed vector increases in the same proportion as the actual wind speed vector the relative wind angle will remain constant. In other words, referring to Figure 5 again, as the wind speed and rotor speed changes the triangle formed by the blade speed, actual wind speed and relative wind speed will maintain exactly the same geometry. The only change will be the size of the triangle.

The second type of wind turbine is the Fixed Speed or Constant Speed design. These turbines are also referred to as "Stall Regulated" turbines. In this design there is a "Cut-In" wind speed. The cut-in speed is a wind speed at which there is enough energy in the wind such that the rotor is able to begin to extract some of this energy from the wind. The cut-in wind speed is dependent upon the actual turbine design, but it is often in the area of 10-12 MPH. Once the wind speed exceeds the cut-in wind speed the rotor RPM remains constant. In actuality there is a slight increase in rotor RPM with wind speed, but it is very small, perhaps on the order of 2% or 3% which, from an aerodynamic perspective is largely insignificant.

We can see from Figure 5 that with the Fixed Speed design the angle of attack will vary with wind speed. For example, the Figure shows a rotor speed of 100 MPH and a wind speed of 20 MPH. This gives an angle of attack off 11.3°. See Equation (5), Angle of Attack = arctan (Wind Speed/Rotor Speed). If the wind speed were to increase to 30 MPH the angle of attack will increase to 16.7°. This happens because the rotor speed remains constant at 100 MPH. Similarly, if the wind speed were to drop below 20 MPH the angle of attack would decrease below 11.3°. So, we see that, for a Fixed Speed rotor, the angle of attack is a function of wind speed. In the real world the windspeed is constantly changing, sometimes on a second to second basis so, in this case, the angle of attack will also change on a second to second basis.

We have already learned that above a certain angle of attack the airfoil will go into stall mode where the lift will decrease, and the drag will increase. For a Fixed Speed rotor, the result is that above a certain wind speed the rotor will experience stall and the power output of the rotor will begin to decrease even as the wind speed continues to increase. This characteristic has important implications for the rotor design of Fixed Speed turbines. For example, airfoils with relatively sharp stall characteristics, such as the NACA 2415 described earlier and the NACA 4412 described in a later section, can cause undesirable structural loading events during an abrupt stall. For Fixed Speed rotors, airfoils with gentle stall characteristics should be selected.

Unlike stall regulated turbines, variable speed turbines continue to increase their power output as the wind speed increases above the rated capacity of the turbine so there must be some method of limiting energy capture at very high wind speeds so as not to damage the generator and drive train. At least for larger turbines, this is commonly done using blade pitch control. The blades can be "Feathered" (pitched into the wind) in order to reduce power capture or to shut down the turbine in excessive winds. In aerodynamic terms turning the blade into the wind reduces the angle of attack, reducing the lift coefficient thereby reducing energy capture from the wind. An added benefit is that blade pitch control can also be used to increase the energy capture by fine tuning the blade pitch at certain wind speeds.

Airfoils Designed for Wind Turbines

Some of the airfoils used on small to medium sized wind turbines during the 1970s and 1980s and even more recently were the NACA 4 digit series airfoils such as the NACA 4412 airfoil shown in Figure 24. The numbers in the "4412" designation have special significance. The first digit describes the maximum camber as a percentage of the chord. The second digit describes the distance of the maximum camber from the leading edge and the last two digits describe the maximum thickness as a percentage of the chord. The 4412 airfoil would therefore be 12% thick.

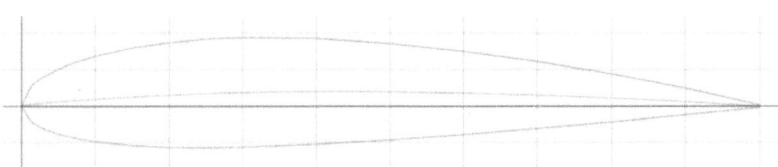

Figure 24 - NACA 4412

Airfoil lift coefficients for the NACA 4412 are shown in Figure 25. The relatively sharp stall at the lowest Re (just over C_L = 1.5) is characteristic of many of the airfoils designed for high Re aircraft use. As already noted, the sharp stall would be detrimental to stall regulated wind turbines because it may lead to high structural loads on the drive train. We will discuss stall regulated wind turbines in a later section but suffice it to say, for the present, that this type of design benefits from airfoils having gentle stall characteristics. The 4412 airfoil is also sensitive to surface roughness so performance would be degraded by fouling of the leading edge by bugs, leading edge tape or wear – all common occurrences for wind turbine rotors.

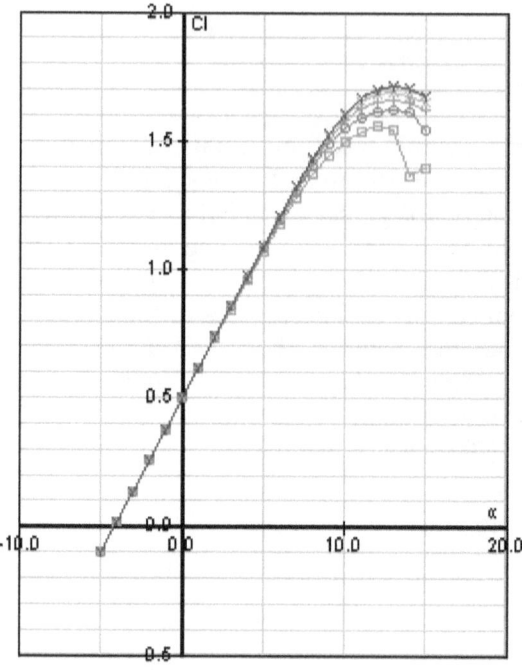

Figure 25 - NACA 4412 Lift Curve

The relatively high lift coefficient and stall characteristics can also result in peak power spikes which can damage the gearbox, generator or other drive train components in high wind and gusty wind conditions, particularly for stall regulated turbines. Because the airfoil was designed for high Re, its performance at low Re may be unpredictable, particularly in the turbulent wind conditions commonly found near the earth's surface where wind turbines are situated.

Beginning in 1984 the United States National Renewable Energy Laboratory (NREL) in a joint effort with Airfoils, Inc., developed a series of special purpose airfoils for horizontal axis wind turbines. The airfoils were designed to exhibit maximum lift coefficients which were relatively insensitive to the effects of roughness. Some of the airfoils were designed for stall regulated rotors with thicker tip airfoils designed to incorporate overspeed control devices. Other features included relatively low maximum lift coefficients which allowed larger diameter rotors for providing better low wind performance. Families of airfoils were developed for small (1-5 meter blades), medium (5 to 10 meter blades) and larger turbines having blades 10 to 25 meters in length. Estimates for annual energy improvement from the use of NREL airfoils is estimated[2] to be 23% to 35% for stall regulated turbines and 8% to 10 % for variable pitch turbines.

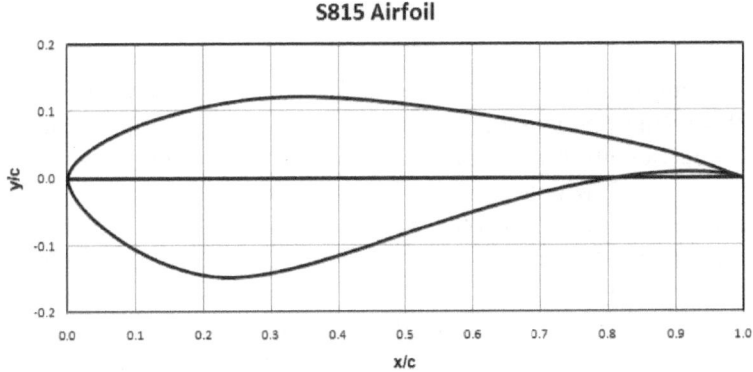

Figure 26 - NREL S815 Airfoil

Figure 26 and Figure 27 show two of the NREL airfoils. The S815 airfoil was designed for the root section of a turbine having a capacity of 100 kW to 400 kW with a blade length of 10 to 15 meters (65' to 100' rotor diameter). The root section is the section of the blade nearest the hub or rotor shaft. The S810 airfoil was designed for the tip section of the blade. The maximum lift coefficient is only 0.9 which is very low compared to airfoils commonly used on aircraft. For example, compare this with the lift coefficient of the NACA 4412 discussed earlier which has a lift coefficient in excess of 1.5. The low lift coefficient allows the use of longer blades thereby increasing swept area and providing better low wind performance than that of an airfoil with a higher lift coefficient. Both of these airfoils tend to be insensitive to surface roughness.

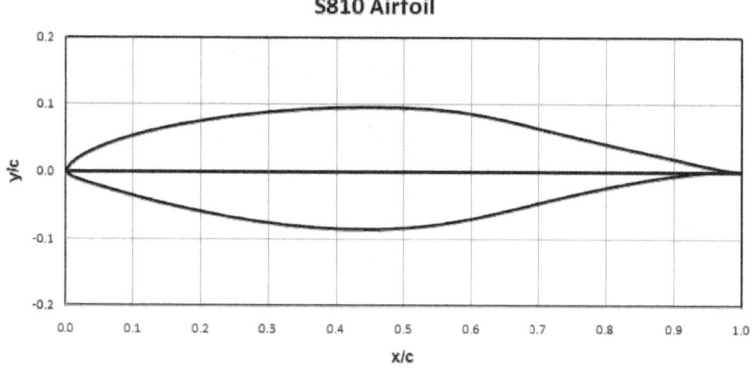

Figure 27 - NREL S810 Airfoil

A lift curve is shown in Figure 28 for the S810 airfoil for Re of 2 million (the open squares). It is interesting to compare the flat lift coefficient above an angle of attack of 10° with that of the NACA 4412 which drops off precipitously above 12°. The moderate reduction in rotor thrust at high angles of attack would result in gentle torque reduction. Note that at high angles of attack, the lift does not change much, however the drag does begin to increase which would result in a reduction in power production. This is also apparent from the L/D graph which shows that above an angle of attack of about 5° the L/D decreases. The lift/drag for the S810 is shown in Figure 29. The L/D reaches a maximum value of about 130 at a 5° angle of attack.

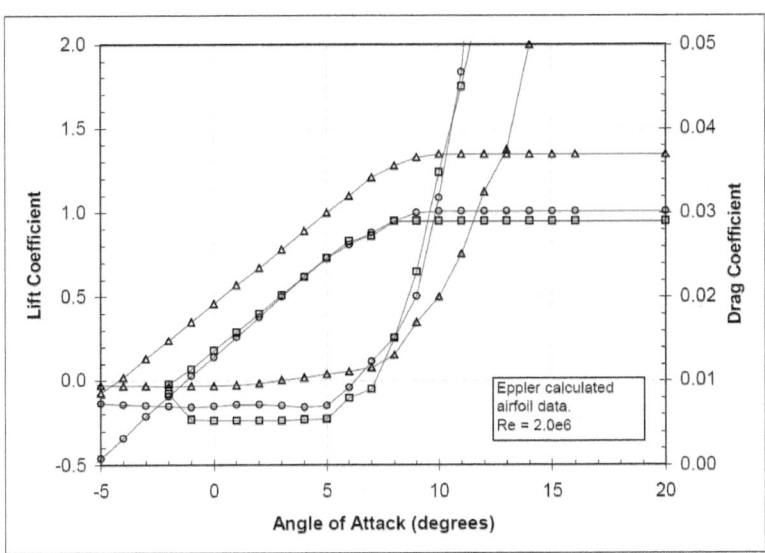

Figure 28 - NREL S810 Airfiol Lift and Drag Coefficients

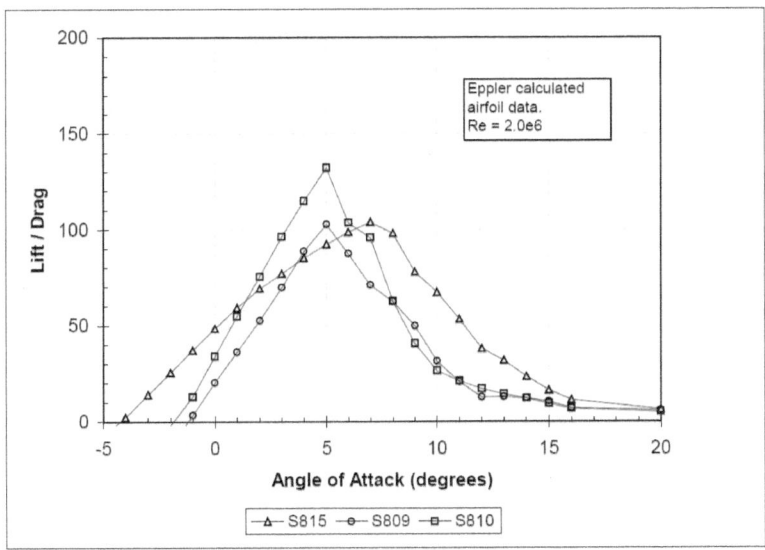

Figure 29 - NREL S810 Lift/Drag

Power Production of a Wind Turbine Rotor

In this section we will look at the factors affecting the power production of a wind turbine rotor and some methods used to calculate the power production.

Airflow in the Vicinity of the Rotor

The airflow in the vicinity of a wind turbine rotor is complex. A typical flow scenario is shown in Figure 30. The velocity U_1 is the wind velocity far upstream of the rotor disk. Velocity U_2 is the velocity far downstream of the rotor and U_{avg} is the velocity at the rotor disk. Note that the wind velocity in the images is designated by the variable "U" but in some cases the variable "V" may also be used interchangeably. The velocity of the wind as it passes through the rotor disk is constant. This may seem counter-intuitive because the rotor presumably extracts energy from the wind so it seems logical that the wind speed on the downwind side of the rotor would be decreased.

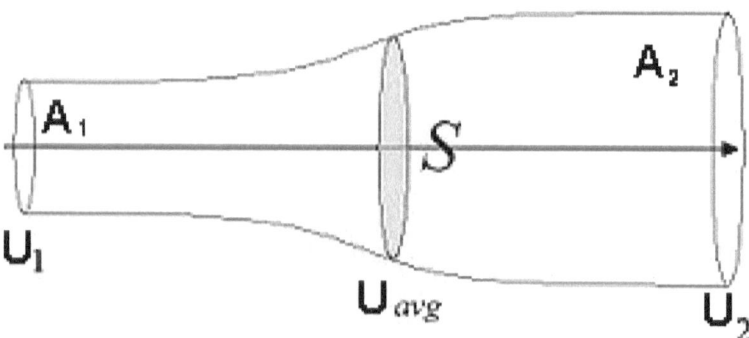

Figure 30 - Airflow near Wind Turbine

The explanation lies in the theory of conservation of mass. If the rotor actually slowed the wind, there would be no place for the approaching air to go. What actually happens is the air pressure changes across the rotor disk. The air pressure on the upstream side of the rotor is greater than atmospheric pressure and the pressure on the downstream side of the rotor is less than atmospheric pressure. Energy is extracted as a result of this pressure drop. This explanation is consistent with the Bernoulli theory which explains the pressure differential existing between the upper and lower surfaces of an airfoil. Recall that the upper surface of an airfoil has lower pressure than the lower surface. The lower surface of a wind turbine blade is the surface that faces the wind so the pressure on this side would be greater than the pressure on the downwind side.

Conservation of mass means that, for a constant density airstream, the product of the area and the velocity must remain constant from upstream of the rotor, through the rotor, to a point downstream of the rotor. The product of velocity and mass is referred to as the **mass flow rate** which is given by:

$$m = \rho A_1 U_1 \qquad (10)$$

where r is the air density, A_1 is the area and U_1 is the velocity. Because of conservation of mass the mass, the flow rate must remain constant for all locations in the moving air stream.

Because the rotor causes a pressure increase on its upwind side some of the air is deflected around the rotor disk. This causes the velocity at the rotor disk to be less than the free stream velocity upwind of the rotor disk. In Figure 30, $U_1 > U_{avg}$ and $A_2 > A_1$. Although the air velocity immediately at the rotor disk, on each side of the rotor, is the same, the air velocity at the rotor disk is an average of the upstream and downstream velocities. The ratio of the reduction in free stream velocity to the velocity at the rotor is called the **axial induction factor** and is given as:

$$a = \frac{U_1 - U_2}{U_1} \qquad (11)$$

where U_1 is the free stream velocity upwind of the rotor and U_2 is the velocity at the rotor. The axial induction factor may be understood simply as a factor giving the reduction in wind velocity at the rotor.

As Figure 30 shows, the upstream area is less than the area of the actuator disk. This is predicted by Equation (10) which would require a larger area at the rotor because of the lower velocity at the rotor.

The difference in pressure across the rotor disk results in a thrust force on the rotor acting normal to the plane of rotation. The thrust force places a bending moment on the blades. This force must be taken into consideration when designing turbine structural components such as the rotor blades, bearings and the tower.

The Betz Limit

If it were possible for a wind turbine rotor to be 100% efficient, it would extract all of the energy from the wind. This is clearly impossible because the rotor would cause the wind speed to drop to zero. In Figure 30, U_{avg} would equal 0 so there would be no wind velocity and therefore no energy for the rotor to extract. We have seen that as a turbine rotor extracts energy from the wind the increase in pressure in front of the rotor disk causes some of the wind to flow around the rotor. This increases the affected area, so in order for the mass flow rate to be constant, the wind velocity must decrease. The more the rotor decreases the wind velocity, the greater the amount of wind deflected around the rotor. Clearly, at some point, the rotor's ability to extract energy from the wind will be offset by its deflecting some of that energy away from itself.

Albert Betz, a German physicist, published the Betz Law in 1919. The law is based on conservation of mass and momentum that we have just reviewed. The Betz Law states that no wind turbine can extract more than 16/27 or 59% of the available energy in the wind.

Wake Rotation

So far, we have assumed a uniform actuator disk with an infinite number of blades with no losses. In reality, wind turbines have a finite number of blades, usually two or three. Modern wind turbines rotate at high enough speed to approach the ideal theory. However, the conservation of momentum theory does not address the torque generated by the blades.

We have seen, in the section on airfoils, that when an airfoil is subjected to a moving air stream such as when a rotor blade is turning in a wind stream, the airfoil pressure distribution can be resolved into both lift and drag which can then be resolved into a thrust force. In the case of a wind turbine rotor the thrust force acts in the plane of rotation of the rotor. This is shown in Figure 19. On a horizontal axis wind turbine, the thrust force results in a moment about the rotor shaft called **Torque**. Torque is a force applied at a distance which tends to produce a rotation about an axis. For example, when you use a long wrench to tighten a nut on a bolt you are applying torque to the nut. In the same fashion, a wind turbine rotor operating in an air stream applies torque to the rotor shaft. The resultant energy that the rotor transmits to the rotor shaft is applied (often through a gearbox) to the generator so it can convert mechanical energy into electrical energy.

In accordance with Newton's laws, for every force there must be an equal and opposite force. The rotating blades must therefore impart rotational energy to the wind. This is called **wake rotation**. The wake rotates in a direction opposite the rotor. An example can be seen in Figure 31 which depicts a wind turbine rotor in a wind tunnel. In this image, the rotor is turning counterclockwise but the wake is rotating clockwise.

Figure 31 - Wake Rotation

Calculating the Power in the Wind

The theoretical power that a wind turbine can extract from the wind is calculated by the following equation:

$$P = \frac{1}{2}\rho A V^3 C_p \qquad (12)$$

where P is power, ρ is air density, A is the swept area of the rotor, V is the wind velocity and C_P is the power coefficient. As we have seen, the maximum power as limited by Betz would be 59% so C_P in this case would have a maximum value of 0.59. One of the most important facts to take away from this equation is that the **power in the wind changes with the cube of the wind speed.** Of course, because of the power coefficient, the actual power which any wind turbine can extract from the wind will be reduced by the power coefficient. Because of this "cube law", as we might call it, an increase in wind speed from 10 MPH to 12.6 MPH would double the available power in the wind.

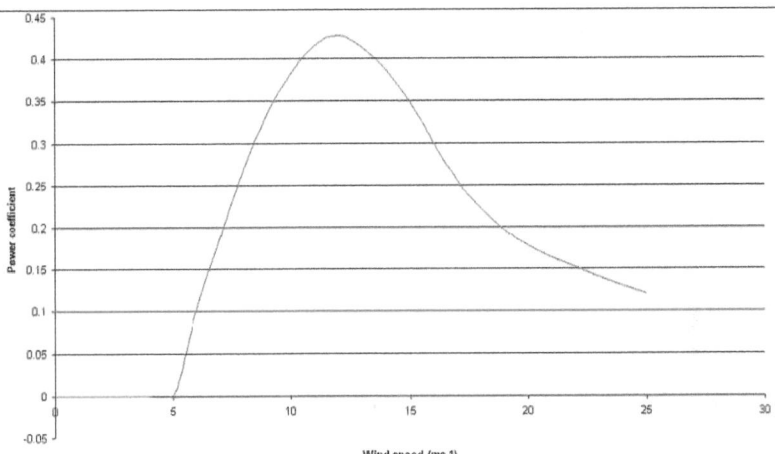

Figure 32 - Rotor Power Coefficient

Power coefficients for a constant speed turbine rotor is shown in Figure 32. The power coefficient curve is shown as a function of wind speed from 5 MPH to 25 MPH. Notice that the power coefficient is not constant. It varies with the speed of the wind. Also notice that the maximum C_P is approximately 0.43, well below the Betz limit. The power begins to drop off at about 12 m/s and continues downward as wind speed increases. This power reduction is typical of a stall regulated rotor. The concept of stall regulated rotors will be discussed in detail further on in this chapter. One way to think about the power coefficient is that it is a measure of the efficiency of a wind turbine rotor in extracting energy from the wind.

Blade Element Momentum Theory

Each portion of the rotor of a horizontal axis wind turbine travels at a different speed. The tip of the blade travels the fastest and the inner section near the hub (rotor shaft) travels more slowly. This is the reason that efficiently designed rotor blades are twisted near the hub. The slower speed of the blade in this area requires an increase in the pitch angle in order to maintain an angle of attack within the operating range of the airfoil and to help to maximize the efficiency (lift over drag) of the airfoil. You may want to review the section discussing the relative wind direction if the reason for this is not clear. This also the reason that inner portions of a rotor blade are wider than the outer portions. According to Equation (8) the lift developed by an airfoil is dependent upon the area of the airfoil and the velocity of the airfoil (or the velocity of the wind at the airfoil location). So, because the velocity of an inner blade section is less than at the tip, the blade will need to have more area (have a longer chord) in order to help make up for the loss in velocity.

It might be useful to compare a rotor blade to the wing of an airplane. On an airplane wing, because all portions of the wing are travelling at the same speed, the design angle of attack of each wing section needs to be the same in order to generate the needed lift. (In actual airplanes, there is usually a slight twist in the wing, but this serves another purpose).

Because of the different velocities, pitches and areas of each section of a rotor blade calculating the power production of the rotor is a bit more complicated than that of a simple airplane wing.

A common method of calculating the power output of a wind turbine rotor is the Blade Element Momentum Theory (BEM). A blade is broken up into a number of radial segments or "elements" and each element is analyzed. It is assumed that the elements do not interact with each other. Because, as we have already noted, different radial sections of a blade travel at different rotational speeds, the velocity and angle of the relative wind varies radially. We have also noted that in order to increase rotor efficiency, most wind turbine blades are tapered and twisted in order to accommodate variations in relative wind velocity and the angle of relative wind.

Equation (12) includes the "A" term for the area of the airfoil section being examined. However, when we use this equation to calculate the power of a horizontal axis wind turbine, we will use the area swept by the blade section under consideration. This area is termed the "annulus" which simply means the area of a donut or ring. This annulus is defined by "dr" in Figure 33. The annulus is the area of the "ring" swept out by the radial dimension dr. The equation also includes a velocity "U" term to calculate the speed at which the blade section is moving relative to the relative wind at the radius of dr. Overall performance is calculated by integration along the blade span.

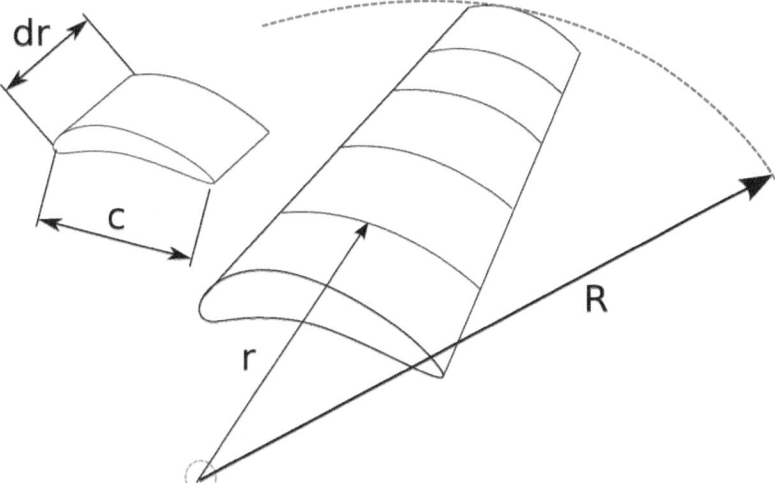

Figure 33 - Blade Element Model

Because of the complexity of the calculations, and the necessity of solving simultaneous equations, they are normally performed using a computer. One software program that was developed by the National Renewable Energy Laboratory is called WT_Perf which uses BEM to predict the performance of a wind turbine rotor. The software calls up an ascii file created by the user which specifies certain rotor parameters such as rotor diameter, rotor RPM, number of blades, number of blade stations, airfoils, the radius, dimensions and blade pitch for each of the stations, range of wind speeds to be analyzed, etc. The program outputs the power in watts and power coefficients for each wind speed. Of course, in order to calculate the actual number of kilowatt hours that a turbine would produce at an actual wind site will require using the information provided by WT_PERF in conjunction with drive train and generator losses and the wind speed information and distribution at a particular site.

Variations exist with regard to the level of complexity of the calculations such as whether or not corrections for wake rotation and/or tip losses are included in the calculations. Tip losses will be discussed in the following paragraphs.

Blade Tip Losses

The difference in pressure between the upper and lower surfaces of an airfoil in an airstream causes a flow from the high pressure to low pressure area at the blade tip. For an airplane wing the high pressure below the wing curves around the end of the wind to the low pressure on the upper wing surface. These vortices can be seen coming off the wingtips of the jet fighter in Figure 34.

For a wind turbine blade, the high pressure on the upwind side of the blade curves over the end of the blade tip to the downwind low-pressure area.

The effect of blade tip losses can be seen as the rotating vortices and spiral coming off the blade tips as in Figure 31. Vortices are tunnels of rotating air. If the vortices are created by large, heavy aircraft the effect is called wake turbulence and the vortices can prove hazardous to small aircraft flying close behind. These vortices represent energy that is wasted. The energy may be sufficient to roll over a small airplane or cause it to lose control if it happens to fly into the tip vortices of a large heavy aircraft. For a wind turbine the effect is to reduce the efficiency of the rotor.

Figure 34 - Wingtip Vortices (Courtesy Wikipedia)

In addition to a loss of efficiency, the turbulence associated with blade tip vortices can cause noise. Various blade tip designs have been created in an effort to reduce tip losses, including square edge, elliptical shapes and curved leading and trailing edges.

Blade Tip Speed Ratio

The blade tip speed ratio is the ratio of the speed of the blade tip to the speed of the wind. The tip speed ratio of a wind turbine blade (TSR) is calculated as:

$$TSR = \frac{Blade\ Tip\ Speed}{Wind\ Speed} \qquad (13)$$

Tip speed ratios for various types of wind turbines are shown in Figure 4.29 as a function of power coefficient. For modern two or three blade turbines tip speed ratios vary between about 4 and 8. Peak C_P occurs between TSRs of 6 to 7. Tip speed ratios should not be confused with tip speeds. Tip speed is simply the speed of the blade tip in MPH or m/s and it is unrelated to the wind speed. For a turbine with a constant RPM rotor the tip speed ratio will vary with wind speed, however the tip speed will be constant. For a variable speed turbine, the tip speed ratio could be nearly constant, but the rotor RPM will vary with wind speed.

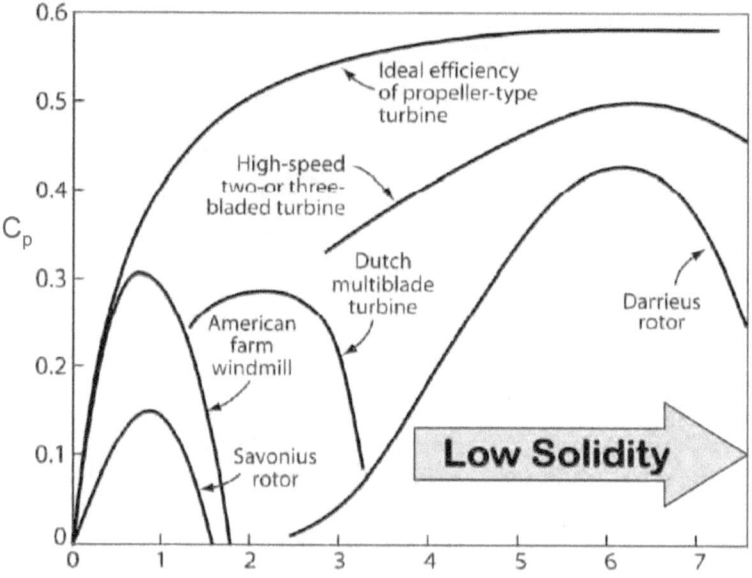

Figure 35 - Tip Speed Ratios of Various Turbine Types

Tip speeds are generally constrained by aerodynamic noise. As the blade velocity increases, sound pressures increase. Noise is generated by turbulence primarily at the blade trailing edge and blade tip. For turbines that are located on land, particularly if the area is inhabited, tip speeds are limited to about 65 m/s (145 MPH). If the turbine is to be installed offshore, higher tip speeds may be permissible. There is no universal rule limiting maximum tip speed for land-based turbines as noise laws or ordinances will vary with location or jurisdiction.

Solidity

The solidity of a wind turbine rotor is the ratio of rotor blade surface area to the swept area and is given by:

$$S = \frac{A_b}{A} \qquad (14)$$

where S is the solidity, A_b is the area of all of the blades and A is the swept area of the rotor. Blade area is the planform not the projected area of the blades as blade twist would cause the projected area to be less than the area defined by the chord length.

Solidity and tip speed ratio are closely related and interdependent. The ideal rotor solidity will vary with the tip speed ratio. High tip speed ratios will require low solidity and low tip speed ratios will require high solidity. An elementary but useful way to think about solidity is that if the rotor speed is high, the blades do not need to be as wide in order to capture the wind passing through the rotor disk. If the blades are turning very slowly, wide blades are necessary in order to prevent the wind from "passing through" the rotor disk without being captured. If blades are too wide for a given tip speed ratio, performance will suffer because the blades will tend to create more turbulence than blades which provide the optimum solidity.

Low solidity has a number of advantages. High rotor speeds tend to have higher power coefficients and greater efficiency. Because the blades are narrower there is less blade mass and less cost of materials, manufacturing and shipping. The higher the rotor RPM, the lower the torque on the rotor shaft and the lower the gearbox ratio resulting in weight and cost reductions. When the rotor is stopped in a high wind there will be less blade area presented to the wind so that rotor and tower loads will be reduced.

High solidity also has disadvantages. The higher speed will result in more acoustic emissions - noise. There may be greater leading-edge erosion thereby requiring more expensive protection. Higher speed can result in greater tip losses. Thinner blades may not be able to resist bending loads as well as wider blades and flexing may be greater possibly creating tower clearance issues.

Number of Blades

One of the first steps in designing a wind turbine rotor is to decide on the number of blades. Almost all large, utility scale, turbines use three blades. Although there has been considerable research on two bladed designs, there have been few large two bladed designs produced. Two bladed designs have, however, achieved some level of popularity in smaller wind turbine designs, particularly in the United States. There are advantages and disadvantages with two bladed turbines as compared to three bladed turbines. One big advantage of the two bladed design is the cost savings of a blade.

When a horizontal axis turbine yaws in response to changing wind direction the rotor will impose a gyroscopic force on the turbine and tower. The amount of force will be dependent on factors such as rotor mass and the yaw rate. Because the mass distribution of a two bladed rotor is linear, the gyroscopic force varies with the blade position during rotation. This can cause severe and even damaging vibration at two cycles per revolution. Because of the more uniform mass distribution of three blade rotors this large pulsating gyroscopic force does not exist. Three bladed rotors may still generate large gyroscopic forces, but they do not generally exhibit the pulsations inherent in two bladed designs. One way to reduce or eliminate the yaw vibration in a two bladed rotor is to decouple the rotor from the turbine rotor shaft by using a teeter bearing much like the common seesaw found in children's playgrounds. This is not new technology as it is commonly used on the tail rotors of helicopters. A full treatment of this subject is beyond the scope of this book, but the reader is advised to thoroughly research the matter if a two bladed design is contemplated. An example of a teeter bearing on a two bladed rotor is shown in Figure 36.

Figure 36 - Teeter Bearing Example

Stall Regulated Wind Turbines

One of the primary concerns of a wind turbine designer is being able to ensure that the turbine is not damaged from excessive loading while running in high winds or that safe operation at high winds is not compromised. We have seen that the force of the wind increases with the square of the wind speed. We have also seen that the energy in the wind increases with the cube of the wind speed. High winds can therefore place very large loads on the turbine and tower.

We have briefly discussed the difference between variable speed turbines and stall regulated turbines. Each of these designs has its advantages and disadvantages. Variable speed turbines, unless they are very small, usually require some sort of blade pitch regulation to prevent overloading the generator and drive train at high wind speeds. The mechanisms required to perform the regulation can be mechanical, electric or hydraulic or some combination of these. Microprocessor controls may be required. These components tend to be expensive and they contain many moving parts subject to wear requiring maintenance. A multitude of parts can also increase the likelihood of failures potentially placing the turbine out of commission until repairs can be made. Very small wind turbines of 10 kW or so, suitable for residential applications, often use simple mechanical blade pitch devices which can be reliable and offer low maintenance. On larger turbines, a big advantage of variable pitch is that it can offer precise control of the blade pitch. This allows the pitch mechanism to not only pitch the blades for shut down in high winds but also to optimize pitch during normal operation.

Stall regulation, on the other hand, is a very simple and robust method of control which has been used for many years on both large and small wind turbines. Stall regulated wind turbines often have fixed pitch blades. This allows the blades to be attached to a rigid hub assembly. This completely eliminates complex and expensive blade pitch systems. Nevertheless, some stall regulated turbines do pitch the blades into feather in order to shut down the turbine in high winds or for maintenance.

The key to stall regulation is maintaining a constant rotor speed independent of wind speed. This is commonly accomplished by designing the drive train and generator so that it provides an adequate load on the rotor. The load must be sufficient to prevent the rotor RPM from increasing even as the wind speed increases to maximum generator capacity. Once the rotor blades go into aerodynamic stall the power output from the blades will begin to decrease. As the wind speed continues to increase, the reduced lift and increased drag from operating in the stall regime will cause the power output of the rotor to continue to diminish with further increases in wind speeds. This is graphically depicted in Figure 37 which is a plot of power versus wind speed. The power curve can be seen increasing with wind speed until it peaks then begins to drop with further increases in wind speed.

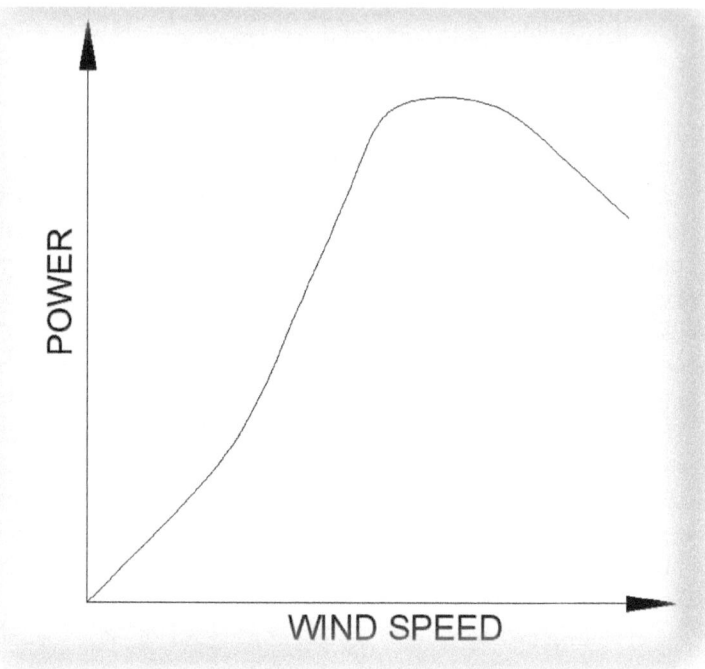

Figure 37 - Stall Regulated Turbine Power Output

Stall regulated wind turbines generally use induction generators because the RPM of an induction generator is essentially constant. The RPM of an induction generator only changes a few percent (~3%) from no load to full load. Because the induction generator is connected to the rotor through a gearbox with a fixed gear ratio the generator provides a constant speed load on the rotor.

Let us consider an actual example. Consider a hypothetical wind turbine with a rotor turning at a constant rotational speed. A plot of the coefficient of lift versus angle of attack is shown in Figure 38. The lift coefficient is labeled on the vertical axis on the left of the graph and the angle of attack is labelled along the bottom. A maximum lift coefficient of about 1.4 is achieved at an angle of attack of about 14°.

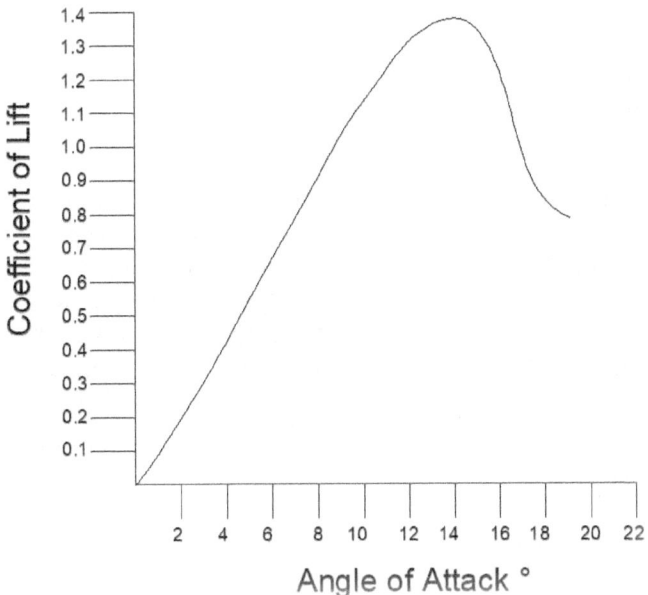

Figure 38 - Airfoil Lift Curve

The blade tip is shown in Figure 39. It is traveling at 100 MPH. The actual wind speed is 21 MPH. We can calculate the angle of attack from Equation (5).

$$Angle = \arctan\left(\frac{21\ MPH}{100\ MPH}\right) = 11.86°$$

Looking at Figure 38 we see that the coefficient of lift at 12° is about 1.3.

Figure 39 - Stall Regulation During Normal Operation

The same blade tip is shown in Figure 40. In this image the wind speed has increased to 30 MPH. Because the rotor turns at constant RPM the rotor speed has not changed so the rotor speed vector does not change. The increase in wind speed results in an increase in the angle of attack to 16.7°. Referring again to Figure 38 we can see that the lift coefficient has dropped to 0.9. Although we have not considered the drag curve, the drag will increase at the higher angle of attack. The circular arrow shown near the trailing edge in Figure 40 is an indication of flow separation and turbulence caused by the excessive angle of attack. The lower lift coefficient and the higher drag at the higher wind speed results in a decrease in rotor output power.

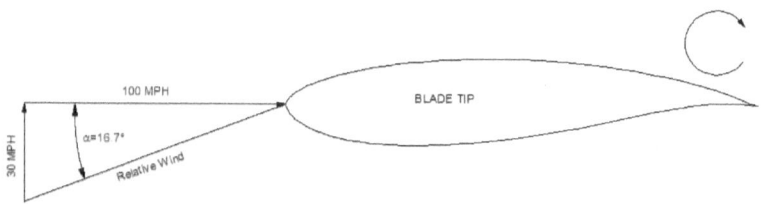

Figure 40 - Airfoil at Stall

This example illustrates that the basic physics governing the relative wind direction has allowed the design of a very simple and potentially inexpensive fixed pitch wind turbine rotor that is essentially self-regulating at all wind speeds. As stated earlier, it is necessary that the generator and drive train be capable of maintaining adequate load on the rotor at high wind speeds so that its RPM cannot increase. If the RPM is allowed to increase with wind speed, the angle of attack would remain below stall and rotor power would increase exponentially with the potential of causing failure of the rotor, drive train and/or generator.

Some of the early attempts at designing stall regulated wind turbines encountered problems because the available airfoils were not designed to have the reliable and gentle stall characteristics essential for stall regulated rotors. This was discussed earlier in the section on airfoils. Airfoils which have high lift coefficients and sharp stall may cause heavy loads on the drive train during high winds and the accompanying gusty wind conditions. The efficient and safe design of a stall regulated rotor requires very careful airfoil selection and blade design.

Although stall regulated wind turbines are capable of regulating their power output at normal operating wind speeds it may be necessary or prudent to shut the turbines down in very high winds in order to limit loads on the turbine and tower. Some control systems are designed to stop the rotor or yaw the turbine out of the wind when the speed is in excess of some value such as 50 or 60 MPH.

Provision must be made to slow or stop the rotor in the case of faults or failures in the utility grid or turbine drive train. For example, if the utility power were to fail in a high wind, the rotor would no longer have a load and it would immediately go into overspeed. Stall regulated turbines are often equipped with mechanical fail-safe brakes on the rotor shaft or high-speed shaft between the gearbox and generator. Rotors are usually also equipped with some aerodynamic means of slowing the rotor such as tip brakes which deploy if the RPM exceeds a certain value. Tip brakes create aerodynamic drag and slow the rotor. An example of tip brakes on a turbine rotor is shown in Figure 41. (The author holds a patent on the tip brake design shown in the image.) Tip brakes may not stop the rotor completely, but they slow it to a very low and safe RPM. In order to ensure that the turbine rotor can be safely stopped or slowed in high winds, all wind turbine designs with induction generators should have at least two fail-safe methods of stopping the rotor. Mechanical, electrical or hydraulic brakes should not take the place of some sort of aerodynamic rotor control such as the tip brakes shown in the image.

Figure 41 - Tip Brakes Deployed

Blade Analysis Example

Introduction

In this section we will look at an example of an actual blade design for a wind turbine. It was noted earlier that calculating the output of a wind turbine rotor is complex and best performed using computer software. A software program entitled WT_Perf was developed by the United States Government National Wind Technology Center, National Renewable Energy Laboratory, Golden, Colorado. The software is a wind turbine performance code which uses Blade Element Momentum Theory (BEM) to calculate the power output of a rotor. The software is available for download from the National Renewable Energy Laboratory, U.S. Department of Energy. The author was able to download the software at no cost. Although our example will make use of this software the following example should nonetheless provide educational value even if the reader does not actually download and run the software.

In the example in this section we will use a three bladed design.

Design Considerations

Wind turbine blade design involves tradeoffs between efficiency and cost of production. The simplest blade design is a rectangular untwisted planform such as the one shown in Figure 42. Such a blade would not be as efficient as a twisted and tapered blade because the inner sections near the hub (blade root) would operate much of the time at high angles of attack and in stall. Additionally, the width of the blade will compromise aerodynamic performance because the tip is wider than it should be, and the root section is narrower than it should be for good aerodynamic performance.

Figure 42 - Simple Rectangular Untwisted Blade

A better design would be to use a tapered and twisted design such as the one shown in Figure 43. This design has a straight trailing edge, but the width varies from the tip to near the root of the blade. The blade is also twisted. This is a very common design choice because it nearly approximates an ideal aerodynamic shape for a wind turbine blade, but it is still not too complex to manufacture economically.

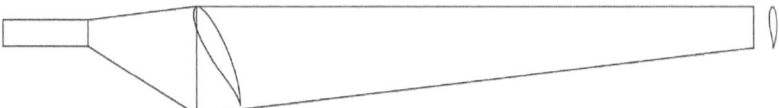

Figure 43 - Simple Tapered and Twisted Blade

This might be a good place to point out that stall regulated rotors, which have gearboxes, may require considerable starting torque to overcome the friction in the gearbox bearings and gears. This becomes particularly important in cold climates where the viscosity of the bearing grease and gearbox oil is high. Under such conditions it may take a considerable amount of wind in order for the blades to have starting torque sufficient to overcome the drive train friction and begin to turn. Once the blades pick up some rotational speed the airfoils will begin to operate more efficiently, and the turbine will quickly come up to running speed. This problem is caused by the very low efficiency of the blades when stopped as the angle of attack will be at or near 90°. Refer to Figure 5 if this is not clear. In such cases, the twisted and tapered blade design, which has a large area near the root section and a substantial blade twist in this location will provide much higher starting torque than the simple rectangular blade. Needless to say, having spent much time and money designing and building a wind turbine it is very frustrating to be standing outside on a cold day with moderate wind speeds watching the turbine rotor just sit there without turning.

If time and money is not a primary consideration, but optimum efficiency and performance are, we could design a blade with the least aerodynamic compromise. This blade might look something like the design shown in Figure 44. Notice the very narrow tip and very wide root section. The blade would also have substantial twist in the root section. In the example that follows we will see that the equations used to calculate blade width and twist will dictate the design shown in Figure 44.

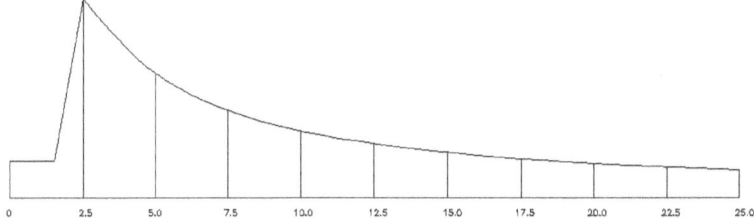

Figure 44 - Optimum Blade Shape

Before selecting a design for our example, it is worth noting that the design of an efficient wind turbine blade is somewhat of an iterative process, often taking several attempts in order to arrive at a reasonable compromise between efficiency and the cost of manufacture.

The Example Design

For the purposes of our example we will use the design shown in Figure 45, which is a compromise between the optimum design shown in Figure 44 and the intermediate design shown in Figure 43. The blade length is 25 feet. We will use a total of 10 stations so the distance between each station is 2.5'. Each of these stations is labeled in Figure 45. When we input the data into WT_Perf we will need to interpolate between the stations so that the data being calculated is the average of the two adjoining stations. These values are shown at the end of the dashed lines between the stations in Figure 45.

The design makes use three airfoils designed by NREL. The airfoils are S819, S820 and S821 and are shown in Figure 46. Airfoil S820 will be used at the blade tip, S819 for the intermediate blade sections and S821 for the blade sections near the blade root.

It is worth pointing out that WT_Perf, or any other BEM modeling software, will require airfoil performance data in order to calculate the power output of the rotor. In a stall regulated rotor, which we know spends much of its time operating at high angles of attack, the airfoil data must include lift and drag coefficients at these large angles. As noted earlier, traditional aircraft airfoil data does not often contain the figures for high angles of attack as this would not have served any purpose in the design of airplane wings. So, finding the necessary airfoil data for the chosen airfoil(s) may prove to be challenging. Fortunately, data does exist for the three airfoils chosen for out example.

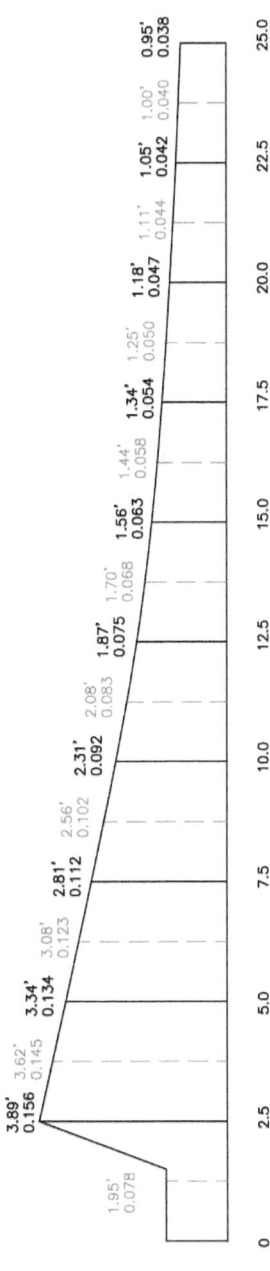

Figure 45 - Blade Design Selected for Example

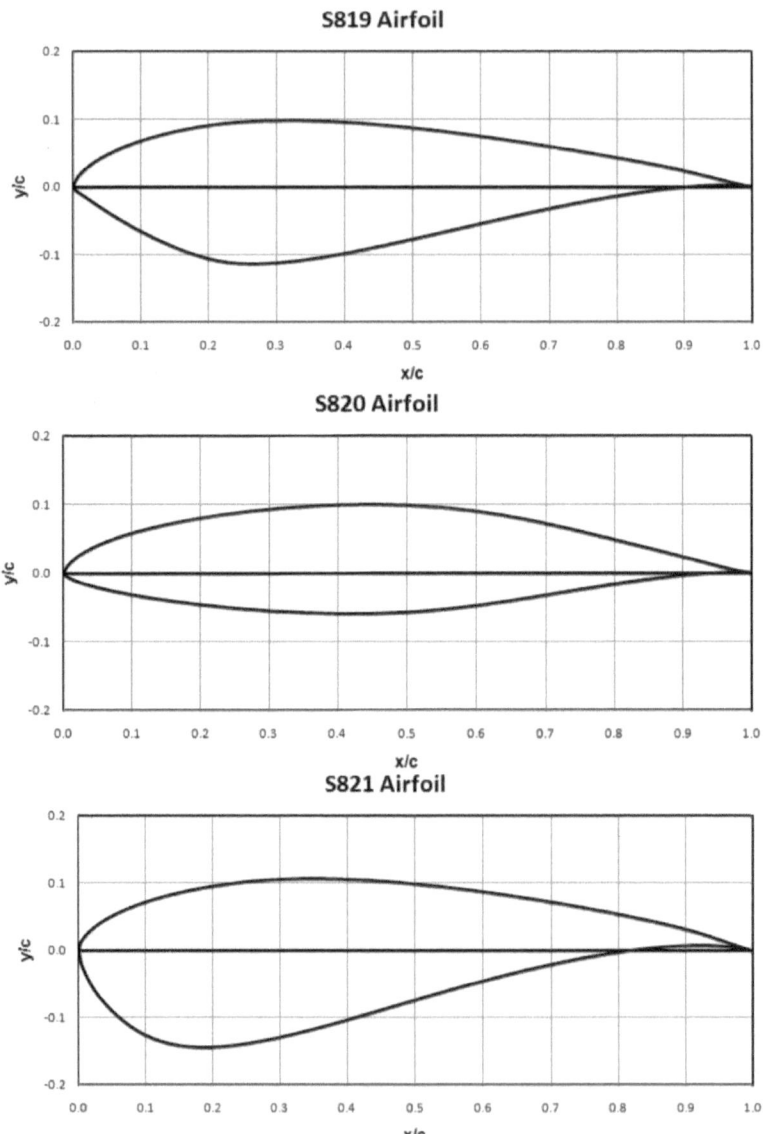

Figure 46 - NREL S819, S820, S821 Airfoils

Calculating the Chord for Each Station

Our first step will be to calculate the blade chord and blade twist for each of the stations. It will be helpful to refer to the blade geometry showing the blade pitch, angle of attack and the angle of the relative wind as shown in Figure 47. It is important to stress that the chord line of an actual wind turbine blade section does not correspond to the plane of rotation. This angle is referred to as the Pitch Angle as shown in the image by Θp. In fact, we will probably want to run WT_Perf using a number of pitch angles in order to understand how the pitch angle affects the power output of the rotor at various wind speeds.

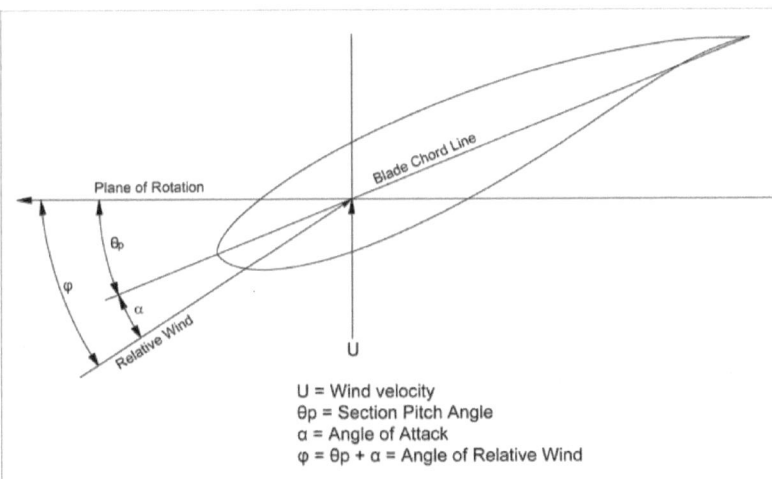

U = Wind velocity
Θp = Section Pitch Angle
α = Angle of Attack
φ = Θp + α = Angle of Relative Wind

Figure 47 - Blade Design Geometry

Blade chord is calculated using the following equation[1]:

$$c = \frac{8\pi r \, Sin\varphi}{3BC_L\lambda_r} \tag{15}$$

---

[1] *Wind Energy Explained: Theory, Design and Application, Second Edition, James Manwell, Jon McGowan and Anthony Rogers, © 2009 John Wiley & Sons, Ltd.*

Where:
r = radius
φ = Relative Wind Angle
B = Number of Blades
C_L = Lift Coefficient
$\lambda r = \lambda \frac{r}{R}$
λ = Tip Speed Ratio
R = rotor radius

For the purposes of our example blade, we will use the following values for the variables:

r will be the radius of the station that we are calculating. For example, r will equal 25.0, 22.5, 20.0, etc.
φ, the relative wind angle is shown in Figure 47.

B, the number of blades will be 3.

C_l, the lift coefficient is dependent on the lift curve of the airfoil that we select. For our example and for purposes of simplicity we will use a lift coefficient of 1.0.

λ is the tip speed ratio. Recall that for a constant speed turbine the tip speed ratio will vary with the wind speed. In our example we will use a tip speed ratio of 7 which is a reasonable value for a three bladed rotor. Refer to Figure 35 for the various tip speed ratios of wind turbines.

R is the rotor radius which is 25.0 feet.

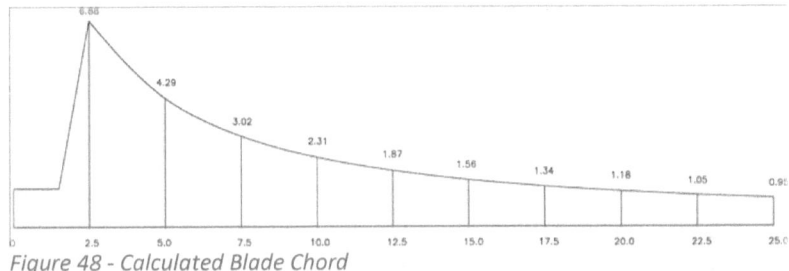
Figure 48 - Calculated Blade Chord

The chord width calculation is performed for each of the stations. For example, the blade chord at the tip of the blade is 0.95 feet. The chord at the widest point near the blade root at station 2.5 feet is 6.88 feet. Results of the calculation are shown in Figure 48.

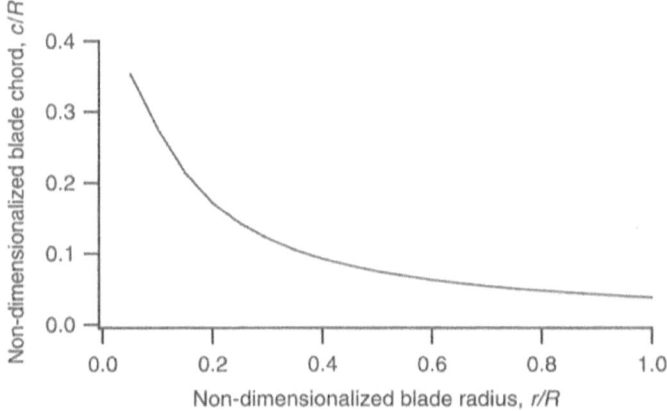

Figure 49 – Blade Chord as a function of Non-dimensional radius.

The blade chord as a function of blade radius is calculated by dividing the chord by the radius. For example, at the tip the calculation would be:

$$C_r = \frac{0.95\ feet}{25\ feet} = 0.038$$

These non-dimensional values are plotted in Figure 49. WT_Perf requires that the chord dimension at each station be input as a function of blade radius so the values at each station will eventually need to be calculated.

The blade chord calculation creates a blade that has a very long chord near the root of the blade. In practice, such a long chord will be expensive to manufacture and to ship, particularly for large blades. Reducing the chord dimensions in this area will not have a great effect on performance because most of the power produced by the blade comes from the outer portions of the blade. Although large inner blade sections may help to start the rotor moving from a stopped position, the actual final design in this area will be a judgement call by the blade designer who will need to balance the benefits versus the costs.

Based on these considerations, for the purposes of our example we will reduce the chord dimensions of the first three stations. We will use the calculated chord widths for Station 10.0' to the tip at station 25.0'. We will use a dimension of 3.89' for station 2.5' and draw a connecting line to the trailing edge at station 10.0. We can then take our chord length dimensions for stations 5.0' and 7.5' from the amended drawing. The resulting blade shape and blade chord dimensions are shown in Figure 45.

Calculating the Blade Twist

We will next calculate the blade twist for each of the stations. Blade twist is the angle between the plane of rotation and the blade chord. Blade twist can be calculated using the following equation[2]:

$$\varphi = \tan^{-1}\left(\frac{2}{3\lambda_r}\right) \qquad (16)$$

Where:
$$\lambda r = \lambda \frac{r}{R}$$

---

[2] Id.

λ = Tip Speed Ratio
r = radius
R = rotor radius
φ = Blade Twist

Note that tan⁻¹ is just another way of stating the arctangent (arctan) of the angle.

A non-dimensional graph of blade twist vs. blade station is shown in Figure 50.

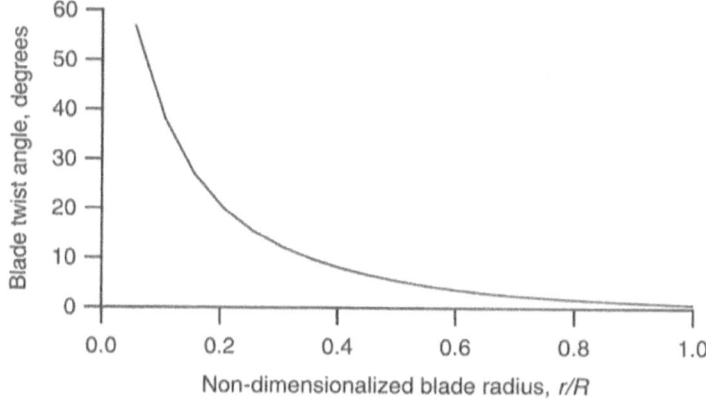

Figure 50 -Blade twist as a function of Non-dimensional radius.

A table of the results of both the blade width and blade twist calculations is shown in Figure 51.

r	r/R	λ r	φ Angle of Relative Wind	Section Pitch	c/R Chord	Chord (Ft)
2.5	0.1	0.70	43.6	36.6	0.275	6.88
5.0	0.2	1.40	25.5	18.5	0.172	4.29
7.5	0.3	2.10	17.6	10.6	0.121	3.02
10.0	0.4	2.80	13.4	6.4	0.092	2.31
12.5	0.5	3.50	10.8	3.8	0.075	1.87
15.0	0.6	4.20	9.0	2.0	0.063	1.56
17.5	0.7	4.90	7.7	0.7	0.054	1.34
20.0	0.8	5.60	6.8	-0.2	0.047	1.18
22.5	0.9	6.30	6.0	-1.0	0.042	1.05
25.0	1	7.00	5.4	-1.6	0.038	0.95

Figure 51 - Blade Width and Pitch Calculation Results.

Note that, in addition to the large chord length near the root (stations 2.5 and 5.0) the blade twist (Section Pitch) is also very large in this area.

Preparing the Data for Input into WT_Perf

We will use the blade dimensions shown in Figure 45 for the WT_Perf data file. As noted earlier we will need to average the data between each station so that the dimensions represent the chord and twist halfway between our stations. We will also need to divide both the stations and chord lengths by the radius so that the dimensions are a function of the radius. The results of our calculations are shown in the table in Figure 52 in the shaded columns.

Modified blade dimensions for WT_Perf

r	r WT_Perf	r/R	r/R	Pitch	Pitch WT_Perf	c/R Chord	c/R Chord WT_Perf
0.0		0.0		0		0	
	1.25		0.05		18.3		0.078
2.5		0.1		36.6		0.156	
	3.75		0.15		27.55		0.145
5.0		0.2		18.5		0.134	
	6.25		0.25		14.55		0.123
7.5		0.3		10.6		0.112	
	8.75		0.35		8.5		0.102
10.0		0.4		6.4		0.092	
	11.25		0.45		5.1		0.084
12.5		0.5		3.8		0.075	
	13.75		0.55		2.9		0.069
15.0		0.6		2		0.063	
	16.25		0.65		1.35		0.059
17.5		0.7		0.7		0.054	
	18.75		0.75		0.25		0.051
20.0		0.8		-0.2		0.047	
	21.25		0.85		-0.6		0.045
22.5		0.9		-1		0.042	
	23.75		0.95		-1.3		0.040
25.0		1.0		-1.6		0.038	

Figure 52 - Modified Blade Dimensions for WT_Perf

WT_Perf is an executable file (WT_Perf.exe) which must be run from a DOS command prompt. The program can be run in Windows by typing the word "command" so that a DOS window will appear. When WT_Perf is run it calls a data file prepared by the user which contains the necessary values for each of the variables needed by WT_Perf in order to perform the computations. Although a complete tutorial for using WT_Perf is beyond the scope of this book, we will cover a few of the more important aspects of the data file.

The WT_Perf data file is an ASCII file (text file) that can easily be created by a simple word processor such as Notepad which is available in Windows. When WT_Perf is run it will output an ASCII text file with the same name as the data file but with the file extension ".oup". This output file will contain the results of the analysis.

The WT_Perf data file for our example blade is shown below.

```
-----  WT_Perf Input File  -----------------------------------------------
WT_Perf BCC 50KW Turbine S819-S821.dat
Compatible with WT_Perf v3.00f
-----  Input Configuration  ---------------------------------------------
False              Echo:              Echo input parameters to "<rootname>.ech"?
False              DimenInp:          Turbine parameters are dimensional?
False              Metric:            Turbine parameters are Metric (MKS vs FPS)?
-----  Model Configuration  ---------------------------------------------
1                  NumSect:           Number of circumferential sectors.
5000               MaxIter:           Max number of iterations for induction factor.
1.0e-6             ATol:              Error tolerance for induction iteration.
1.0e-6             SWTol:             Error tolerance for skewed-wake iteration.
-----  Algorithm Configuration  -----------------------------------------
True               TipLoss:           Use the Prandtl tip-loss model?
False              HubLoss:           Use the Prandtl hub-loss model?
True               Swirl:             Include Swirl effects?
True               SkewWake:          Apply skewed-wake correction?
True               AdvBrake:          Use the advanced brake-state model?
True               IndProp:           Use PROP-PC instead of PROPX induction algorithm?
False              AIDrag:            Use the drag term in the axial induction calculation?
False              TIDrag:            Use the drag term in the tangential induction calculation?
-----  Turbine Data  ----------------------------------------------------
3                  NumBlade:          Number of blades.
25                 RotorRad:          Rotor radius [length].
0.0                HubRad:            Hub radius [length or div by radius].
0.0                PreCone:           Precone angle, positive downwind [deg].
0.0                Tilt:              Shaft tilt [deg].
0.0                Yaw:               Yaw error [deg].
3.0                HubHt:             Hub height [length or div by radius].
10                 NumSeg:            Number of blade segments (entire rotor radius).
RElm    Twist    Chord    AFfile    PrntElem
0.05    18.3     0.078    3         TRUE
0.15    27.5     0.145    3         TRUE
0.25    14.5     0.123    3         TRUE
0.35    8.5      0.102    3         TRUE
0.45    5.1      0.084    2         TRUE
0.55    2.9      0.069    2         TRUE
0.65    1.3      0.059    2         TRUE
0.75    0.2      0.051    2         TRUE
0.85    -0.6     0.045    1         TRUE
0.95    -1.3     0.040    1         TRUE
-----  Aerodynamic Data  ------------------------------------------------
0.00238            Rho:               Air density [mass/volume].
1.576e-4           KinVisc:           Kinematic air viscosity
0.0                ShearExp:          Wind shear exponent (1/7 law = 0.143).
False              UseCm:             Are Cm data included in the airfoil tables?
3                  NumAF:             Number of airfoil files.
"airfoils/S820.dat"    AF_File:       List of NumAF airfoil files.
"airfoils/S819.dat"
"airfoils/S821.dat"
-----  I/O Settings  ----------------------------------------------------
True               TabDel:            Make output tab-delimited (fixed-width otherwise).
False              KFact:             Output dimensional parameters in K (e.g., kN instead on N)
False              WriteBED:          Write out blade element data to "<rootname>.bed"?
False              InputTSR:          Input speeds as TSRs?
"mph"              SpdUnits:          Wind-speed units (mps, fps, mph).
-----  Combined-Case Analysis  ------------------------------------------
0                  NumCases:          Number of cases to run.  Enter zero for parametric analysis.
WS or TSR    RotSpd    Pitch          Remove following block of lines if NumCases is zero.
-----  Parametric Analysis (Ignored if NumCases > 0 )  ------------------
3                  ParRow:            Row parameter    (1-rpm, 2-pitch, 3-tsr/speed).
2                  ParCol:            Column parameter (1-rpm, 2-pitch, 3-tsr/speed).
```

1	ParTab:	Table parameter	(1-rpm, 2-pitch, 3-tsr/speed).
True	OutPwr:	Request output of rotor power?	
True	OutCp:	Request output of Cp?	
False	OutTrq:	Request output of shaft torque?	
False	OutFlp:	Request output of flap bending moment?	
False	OutThr:	Request output of rotor thrust?	
-1, 5, 1	PitSt, PitEnd, PitDel:	First, last, delta blade pitch (deg).	
40, 80, 5	OmgSt, OmgEnd, OmgDel:	First, last, delta rotor speed (rpm).	
8, 40, 1	SpdSt, SpdEnd, SpdDel:	First, last, delta wind speeds.	

Some of the settings in this file are advanced and beyond the scope of this book but using the defaults will still result in quite usable results.

Next, we will consider the user inputs that will be required in order to analyze our rotor. It should be pointed out that the data file formatting is critical as the executable file reads the data file line by line so the user must be exceedingly careful in editing this file not to add any lines or carriage returns. It is best to keep an original of the file in a safe location and only edit copies of the original file. If the formatting of the file is inadvertently altered the executable will not run and the user will receive little to no guidance as to what went wrong. If an original data file is kept it is always possible to go back to the original file, make another copy, and try again.

Looking at the data file, the second line can be edited to show the name of the project. In our example we called it "WT_Perf BCC 50KW Turbine S819-S821.dat". This heading will be printed at the top of the output file.

The section entitled "Turbine Data" will need to be edited as follows:

Number of Blades = 3
Rotor Radius = 25
Number of blade segments = 10. These are our stations.

The next ten lines contain the 10 stations where we will need to input the station radius, blade twist and chord. These begin with the station nearest the hub (station 0.05) and proceed outward along the blade. Recall that the stations are a fraction of the blade radius, so they are numbered in 0.1 increments from 0.05 to 0.95.

The Twist column lists the blade twist in degrees. The Chord column lists the chord length as a function of blade radius.

The next column titled "AFfile" is a reference to the airfoil file. Recall that in our sample blade we decided to use three airfoils: S820, S819 and S21. The number in the AFfile column is the number of the airfoil file. The actual names of the airfoil files are listed in the next section "Aerodynamic Data". You can see from the ten stations that specific airfoils are assigned to specific stations. For example, S820, which is airfoil file number 1 is assigned to the two tip stations 0.85 and 0.95. The airfoil file contains the lift and drag coefficients for the airfoil at different angles of attack. A copy of an airfoil file is shown in the appendix.

Note that in the Aerodynamic Data section it is necessary to specify the number of airfoil files, in this case 3. Following that are the names of the three airfoil files beginning with "S820.dat".

The remainder of the input file contains settings for the data that is output. Many of the defaults can be accepted. The important output settings are contained in the last three lines of the data file. The first of these contains three comma separated numbers in the first column. This sets how many blade pitch angles are to be calculated. In our data file we have chosen to begin the analysis with a blade pitch setting of -1° through 5° in increments of 1°. This means that the output file will contain six results based on blade pitch: -1°. 0°, 1° through 5°.

The next to last line allows the user to specify the number of rotor speeds (RPM) to be analyzed. In our file we will begin the analysis at 40 RPM and increment RPMs by 5 until 80 RPM is reached. Recall that this analysis is for a constant speed rotor so the calculation results will allow us to select the best RPM for our rotor. As the generator RPM will be fixed by the number of generator poles (usually nominal 1,800 RPM for a 4 pole generator or 1,200 RPM for a 6 pole generator) the rotor RPM that we finally select will tell us what gearbox gear ratio we will need for the turbine.

The last line in the data file allows us to specify the range of wind speed for analysis. In our data file we have decided to begin the analysis at 8 MPH and increase the speed by 1 MPH increments up to 40 MPH. This will give us a large number of data points so that we can see at what point the rotor goes into stall so that we can be sure that our selected generator capacity is adequate.

Because we have specified a relatively large number of wind speeds and RPM values our output file will be large. For purposes of this example the results of only a single RPM have been included. Figure 53 shows the rotor power output at 60 RPM for wind speeds 8 to 40 MPH. Power output is shown in watts so it will be necessary to divide the numbers by 1,000 for kW which is a more commonly used measure for turbines of this capacity. The columns are arranged by blade pitch from left to right with -1° on the left and 5° on the right.

```
Results generated by WT_Perf (v3.10, 16-Dec-2004) for input file "Bladel.wtp".
Generated on 05-Apr-2020 at 09:05:12.
Input file title:
   WT_Perf BCC 50KW Turbine S819-S821.dat

Power (W) for Omega = 75 rpm.

                    Pitch (deg)
WindSp   -1.000     0.000      1.000      2.000      3.000      4.000      5.000
(mph)
 8.000  -1949.280   -789.772   -1237.609    14.721    -623.590   -229.789   -50.038    -122.540    -436.217
 9.000   -789.772    691.172    1495.644   686.133    2114.903   976.179    1193.680   1093.466    788.382
10.000    691.172   2481.232   3345.989   1495.644   2114.903   2521.199   2670.452   2472.253   2105.566
11.000   2481.232   4608.347   5520.408   3345.989   3971.941   4387.235   4530.527   4305.564   3634.839
12.000   4608.347   7038.766   8014.198   5520.408   6209.963   6635.943   6734.412   6440.789   5685.577
13.000   7038.766   9880.931  10836.502   8014.198   8762.519   9241.357   9299.995   8893.141   8018.931
14.000   9880.931  13638.039  14199.750  10836.502  11655.618  12144.180  12186.231  11706.493  10647.929
15.000  13638.039  17968.416  18659.957  14199.750  14898.171  15368.025  15364.267  14825.634  13649.729
16.000  17968.416  22292.307  23299.834  18659.957  18944.088  18962.889  18858.566  18221.189  16995.971
17.000  22292.307  27122.078  28137.855  23299.834  23802.871  23421.299  22692.994  21932.605  20586.768
18.000  27122.078  32664.285  33302.230  28137.855  28710.553  28448.916  27377.125  25980.926  24489.229
19.000  32664.285  38472.105  38917.906  33302.230  33938.672  33617.461  32582.439  30788.880  28758.754
20.000  38472.105  44128.445  44885.336  38917.906  38520.395  39188.910  38084.176  36158.496  33656.219
21.000  44128.445  50044.965  50399.020  44885.336  44521.098  43486.348  43942.082  41938.582  39225.594
22.000  50044.965  55572.348  56315.465  50399.020  50726.855  49618.477  48202.949  47908.109  45243.785
23.000  55572.348  60266.727  61831.715  56315.465  56432.305  50825.965  54133.504  52802.227  51360.176
24.000  60266.727  63497.000  66537.055  61831.715  62254.484  61946.480  60744.398  58197.859  56762.375
25.000  63497.000  59721.914  67228.234  66537.055  67743.594  67820.109  67045.484  64956.141  62438.539
26.000  59721.914  55078.969  63663.277  67228.234  72564.984  73376.766  73071.781  71578.250  68733.766
27.000  55078.969  52136.520  60841.547  63663.277  71603.352  78297.695  78644.797  77881.000  75652.812
28.000  52136.520  49185.746  59590.301  60841.547  69132.898  76880.094  83463.203  83704.414  82394.641
29.000  49185.746  45234.941  55127.090  59590.301  67367.797  75155.539  82040.773  88220.891  88265.180
30.000  45234.941  40799.203  52693.156  55127.090  66164.359  72835.016  80697.617  87607.172  92989.797
31.000  40799.203  36390.762  48580.082  52693.156  62030.680  72438.070  77989.867  85869.078  92983.312
32.000  36390.762  29704.281  44913.734  48580.082  60260.727  70995.406  79574.008  83524.305  90864.344
33.000  29704.281  24828.930  40635.922  44913.734  56564.531  67280.555  76470.891  85386.227  89978.008
34.000  24828.930  20240.350  34493.234  40635.922  53076.305  64609.559  74986.352  83405.125  90758.227
35.000  20240.350  13531.717  29737.027  34493.234  49418.242  60702.633  71774.727  81036.242  91396.672
36.000  13531.717   8182.817  25314.373  29737.027  45352.184  57841.742  69324.977  81231.227  87401.414
37.000   8182.817   3133.400  19432.021  25314.373  39873.762  54365.652  66536.758  76795.539  88090.695
38.000   3133.400    350.310  14142.535  19432.021  35073.250  50516.008  62987.137  74217.719  86186.633
39.000    350.310  -6219.061   9269.895  14142.535  30742.719  45716.383  59678.938  72617.727  82138.086
40.000  -6219.061             9269.895   9269.895  25748.611  40837.449  56713.523  68591.516  80292.156
```

Figure 53 - WT_Perf Power output for 75 RPM at various wind speeds.

The results for the 75 RPM -1° pitch column have been plotted in Figure 54 so that the results can be more easily visualized. WT_Perf also outputted the Power Coefficient (C_p) and this is also plotted. Notice that the power output has a sharp peak 63.5 kW at a wind speed of 25 MPH then drops off precipitously. This is an excellent example of stall regulation. By the time the wind gets to 40 MPH the rotor is not producing power. It is worth keeping in mind, however. That the lift and drag data for airfoils may not be reliable at high angles of attack so this area or the graph should be relied upon with caution. Also notice that the power coefficient peaks at 0.48 at 20 MPH. This is well below the Betz limit of 0.59.

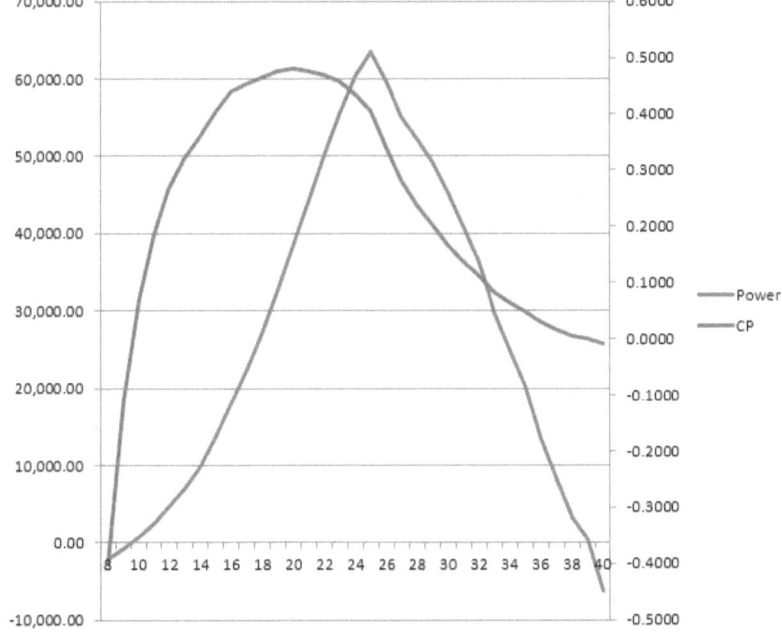

Figure 54 - Calculated Rotor Power for 75 RPM

A thorough analysis of the rotor power would require graphing the results of different pitch angles and RPMs. Although it is beyond the scope of this book, it would be important to consider these results in light of the wind regime at the location at which the turbine would be installed. For example, choosing a configuration that optimized rotor efficiency at a 30 MPH wind speed over a higher power coefficient at lower wind speed might not make sense if the wind regime at the proposed location averaged a much lower wind speed. In other words, it is necessary to consider the wind speed distribution at a particular location in order to make an informed judgment regarding the rotor design.

Other things to consider are the additional noise of a higher tip speed (higher RPM), proposed generator capacity and the cost of the drive train components.

Appendix

Here is a sample copy of the airfoil file used by WT_Perf. Note that the airfoil data files need to be located in a subdirectory where the WT_Perf executable is located. For example, if the WT_Perf executable is located in C:\BLADE ONE, the airfoil data file must be located in C:\BLADE ONE\Airfoils\. This is where WT_Perf will look for the airfoil data.

It should also be pointed out that the input data file always has a file extension of ".wtp" and this file must be located in the same directory as the WT_Perf executable. When WT_Perf is run it will create an output file with the same name as the input file but with the extension ".oup" and this file will be written to the same directory in which WT_Perf is located.

The airfoil file contains information on the airfoil being used in the calculation. Most of the file is devoted to describing the lift and drag of the airfoil as a function of the angle of attack.

```
AeroDyn airfoil file.   Compatible with AeroDyn v13.0.
S819 Airfoil, OSU data at Re=1.25 Million, Clean roughness
NREL/S833 Appendix B, Viterna used aspect ratio=11
   1                Number of airfoil tables in this file
   9.0e9            Table ID parameter (Reynolds number in milllions).
   12.0             Stall angle (deg)
   -2.3             Zero lift angle of attack (deg)
   7.12499          Cn slope for zero lift (dimensionless)
   1.9408           Cn at stall value for positive angle of attack
   -0.8000          Cn at stall value for negative angle of attack
   -0.5000          Angle of attack for minimum CD (deg)
   0.008            Minimum CD value
  -180.00     .000      .1748    .0000
  -170.00     .230      .2116    .4000
  -160.00     .460      .3172    .1018
  -150.00     .494      .4784    .1333
  -140.00     .510      .6743    .1727
  -130.00     .486      .8799    .2132
  -120.00     .415     1.0684    .2498
  -110.00     .302     1.2148    .2779
  -100.00     .159     1.2989    .2933
   -90.00     .000     1.3080    .2936
   -80.00    -.159     1.2989    .2933
   -70.00    -.302     1.2148    .2779
   -60.00    -.415     1.0684    .2498
```

-50.00	-.486	.8799	.2132
-40.00	-.510	.6743	.1727
-30.00	-.494	.4784	.1333
-20.10	-.560	.3027	.0612
-18.10	-.670	.3069	.0904
-16.10	-.790	.1928	.0293
-14.20	-.840	.0898	-.0090
-12.20	-.700	.0553	-.0045
-10.10	-.630	.0390	-.0044
-8.20	-.560	.0233	-.0051
-6.10	-.640	.0131	.0018
-3.86	.000	.0084	-.0216
-3.04	.010	.0080	-.0216
-2.00	.200	.0090	-.0282
-1.00	.140	.0090	-.0282
0.00	.240	.0092	-.0346
1.00	.350	.0100	-.0346
2.00	.450	.0100	-.0405
3.00	.550	.0120	-.0405
4.00	.670	.0160	-.0455
5.00	.770	.0110	-.0455
6.00	.980	.0122	-.0455
7.00	.900	.0146	-.0507
8.00	1.080	.0130	-.0404
9.00	1.140	.0180	-.0321
10.00	1.200	.0210	-.0281
12.10	1.200	.0369	-.0284
13.20	1.200	.0509	-.0322
14.20	1.010	.0648	-.0361
15.30	1.020	.0776	-.0363
16.30	1.000	.0917	-.0393
17.10	.940	.0994	-.0398
18.10	.850	.2306	-.0983
19.10	.700	.3142	-.1242
20.10	.660	.3186	-.1155
30.00	.705	.4784	-.2459
40.00	.729	.6743	-.2813
50.00	.694	.8799	-.3134
60.00	.593	1.0684	-.3388
70.00	.432	1.2148	-.3557
80.00	.227	1.2989	-.3630
90.00	.000	1.3080	-.3604
100.00	-.159	1.2989	-.3600
110.00	-.302	1.2148	-.3446
120.00	-.415	1.0684	-.3166
130.00	-.486	.8799	-.2800
140.00	-.510	.6743	-.2394
150.00	-.494	.4784	-.2001
160.00	-.460	.3172	-.1685
170.00	-.230	.2116	-.5000
180.00	.000	.1748	.0000

EOT

INDEX

Air density, 14
Air Density slugs, 14
Airfoil, 18
Airfoil offsets, 21
Airfoil pressure distribution, 61
Angle of attack, 22
Annulus, 65
Axial Induction Factor, 59
BEM, 64
Bernoulli Principle, 28
Betz Limit, 60
Blade chord, 84
Blade Element Momentum Theory, 64
Blade pitch, 72
Blade tip vortices, 66
Blade twist, 87
Blades, number, 70
Boundary layer, 37
Camber line, 22
Chord line, 20
Clark Y Airfoil, 20
Conservation of mass, 58
CP, 62
Cube law, 63
Density, 13
Drag, 42
Drag coefficient, 44
exponent, 16
Flettner Rotor, 31
Flow around a Cylinder, 28
Gyroscopic force, 70
Kinematic viscosity, 36
Laminar flow, 39
Leading edge, 19
Lift, 42
Lift and Drag, 28
Lift coefficient, 44
Low Pressure, 30
Mass Flow Rate, 59
Momentum Theory, 34
NACA airfoil, 44
Pitch angle, 26
Power coefficient, 62
Pressure, 15
Relative wind, 22
Reynolds number, 34
Rotating cylinder, 30
S819, 81
S820, 81
S821, 81
Shear stress, 36
Solidity, 69
Stall regulated turbines, 72
Streamlines, 29
Teeter bearing, 70
Tip brakes, 77
Tip speed ratio, 67
Trailing edge, 19
TSR, 67
Turbulent flow, 39
Variable speed turbines, 72
Vectors, 24
Viscosity, 35
Wake rotation, 61
Wind velocity, 16
WT_Perf, 65, 78
WT_Perf data file, 90
WT_Perf output file, 93

Yaw, 70

www.ingramcontent.com/pod-product-compliance
Lightning Source LLC
Chambersburg PA
CBHW021450210526
45463CB00002B/719